Lab Ref

VOLUME 2

A Handbook of Recipes, Reagents, and
Other Reference Tools for Use at the Bench

Lab Ref

VOLUME 2

A Handbook of Recipes, Reagents, and
Other Reference Tools for Use at the Bench

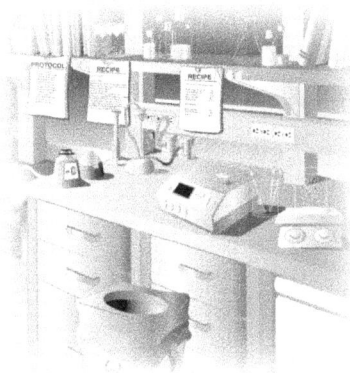

EDITED BY

Albert S. Mellick

Linda Rodgers

Cold Spring Harbor Laboratory

http://www.cshprotocols.org

COLD SPRING HARBOR LABORATORY PRESS
Cold Spring Harbor, New York

Lab Ref
Volume 2
A Handbook of Recipes, Reagents, and Other Reference Tools for Use at the Bench

Publisher	John Inglis
Acquisition Editor	Jan Argentine
Developmental Director	Jan Argentine
Developmental Editor	David Crotty
Project Editor	Mary Cozza
Permissions Coordinator	Carol Brown
Production Editor	Rena Steuer
Desktop Editor	Susan Schaefer
Production Manager	Denise Weiss
Cover Artist	Jim Duffy
Cover Designer	Ed Atkeson

Library of Congress Cataloging-in-Publication Data

Lab ref: a handbook of recipes, reagents, and other reference tools for use at the bench / edited by Jane Roskams and Linda Rodgers
     p. cm.
  Includes index.
  ISBN 978-1-621822-12-7 (pbk.)
    1. Biochemistry--Laboratory manuals. I. Roskams, Jane. II. Rodgers, Linda.
QP519 .L197 2002
572'.078--dc21

2001058104

Contents

A Note from the Publisher

In recent years, Cold Spring Harbor Laboratory Press has published many highly regarded laboratory manuals on a wide range of topics in molecular and cellular biology. These books were carefully designed to address the needs of people who work at the bench. With the same aim, we published a book in 2002 as a companion to a laboratory's collection of manuals. The first *Lab Ref* volume compiled recipes and reference data drawn from a selection of our manuals and was intended to save time and spare frustration.

In the same spirit, *Lab Ref* Volume 2 again assembles in one place a new selection of reference information from a diversity of our published manuals, presented in a format that should maximize the volume's value in a crowded laboratory environment.

We thank the authors and editors of the laboratory manuals who provided source material for this volume as well as Albert S. Mellick, Linda Rodgers, Maarten Hoek, Mary Cozza, Rena Steuer, and Susan Schaefer for the diligence and skill with which they organized and edited the information.

The material in this book can be found in Cold Spring Harbor Laboratory Press manuals and is cross-referenced as coded below in shaded ovals.

Lab Manuals Key

ARA	=	*Arabidopsis*: A Laboratory Manual
IND	=	Imaging in Neuroscience and Development: A Laboratory Manual
LCI	=	Live Cell Imaging: A Laboratory Manual
MC3	=	Molecular Cloning, 3rd Edition
MICRO	=	DNA Microarrays
MME3	=	Manipulating the Mouse Embryo: A Laboratory Manual, 3rd Edition
PCR2	=	PCR Primer: A Laboratory Manual, 2nd Edition
PPI2	=	Protein–Protein Interactions, 2nd Edition
PROT	=	Proteins and Proteomics: A Laboratory Manual
PUR	=	Purifying Proteins for Proteomics: A Laboratory Manual
RNAi	=	RNAi: A Guide to Gene Silencing

Introduction

*L*ike the first volume of *Lab Ref*, *Lab Ref* Volume 2 is organized to allow easy access to the formulae for commonly used laboratory reagents and to facilitate robust laboratory access to information in "the Internet age." In this edition, we have included specialized reagents for protocols not contained in *Lab Ref* such as quantitative nucleic acid analysis, RNA silencing, and imaging.

This book is designed to be used as a handy benchtop source of recipes for reagents needed for protocols described in Cold Spring Harbor Laboratory Press manuals. It is not the intention of the editors to include any buffers or recipes from the first volume of *Lab Ref*. Each recipe is cross-referenced to the manual of origin. Please refer to the Lab Manuals Key listed on the facing page. Reagents, information, and recipes are organized by usage with a focus that begins with quantitative nucleic analysis and ends with more specialized species-specific reagents. Within each category, the reagents are categorized. Also included is a selection of commonly used background information with reference tables and charts.

Due to space constraints, we have not included every recipe from each manual but have instead selected those that are most commonly used (and borrowed by others) by laboratories using techniques covered in these sections.

Here is a brief description of the material and its organization.

Section I

This section comprises recipes for reagents and stock solutions used in quantitative nucleic acid analysis. Areas covered include purification and preparation of material as well as reagents used in analysis. The recipes are useful for analysis of nucleic acid isolated from bacterial, plant, or mammalian tissues. Recently, the explosion in array-based technologies has revolutionized much of molecular biology, thus this section's particular focus is on preparation of material for array analysis. There is less emphasis on quantitative PCR analysis because many commonly used commercial reagents (e.g., SYBR Green I, ABI) are often preferred.

Section 2

RNAi is a relatively new field in molecular biology, having emerged in the last 10 years as a response to observations in *C. elegans* of dsRNA sequence-specific silencing. Since then, small regulatory RNAs have been identified as a normal part of regulating cell biology with specific regulatory enzymes. The section details specialized reagents used in the preparation of dsRNA for gene silencing activity in *C. elegans, D. melanogaster,* and *Arabidopsis,* as well as in mammalian tissues and embryos. Also included are buffers used in the isolation, preparation, and preliminary analysis of these small regulatory RNAs.

Section 3

Techniques in imaging have advanced significantly in the last few years with the notable widespread use of green fluorescent protein (GFP). Other fluorescent labels are now also in use. This section provides recipes suitable for the fixation and preparation of tissues before imaging as well as reagents for staging tissues for examination of cells and intracellular structures such as the cytoskeleton.

Section 4

This section contains reagents used in examination of proteins and protein–protein interactions. Included are quantitative methods, purification, and antibody studies not included in the first volume of *Lab Ref.*

Section 5

This section contains buffers and reagents used in plant studies (specifically, *Arabidopsis*) not covered in previous sections. Areas include culturing conditions, preparation of tissues for histochemistry, and isolation of nucleic acid.

Section 6

Recipes covering buffers and reagents used in studies on the mouse will be found in this section. Areas covered include in vitro culturing of embryonic stem cells and preparation of tissues for cryostorage.

Section 7

This section provides directions for the optimal long- and short-term storage of DNA as well as bacterial, mammalian, and yeast cells.

Section 8

The final section of the book contains formulae, conversion tables, information tables, nomenclature guidelines, useful World Wide Web sites, and background information. It is designed to be used as a reference section for commonly needed facts.

Quantitative Nucleic Acid Analysis

Quantitative Nucleic Acid Analysis

This section of the manual contains instructions for preparing solutions that are commonly used for the preparation and quantitation of nucleic acids, with an emphasis on protocols for global array analysis. Additional information regarding specific applications and uses of these reagents can be found in the cross references to Cold Spring Harbor Laboratory Press manuals (see key on p. viii) that are designated following each recipe.

A specific focus is on isolation of nucleic acids for use in array analysis and preparation of arrays. Most reagents for polymerase chain reaction (PCR) analysis and large-scale nucleic acid preparations are commercially available (e.g., QIAGEN, ABI, etc.), so buffers in use for specific applications (isolation of polysomal RNA) are included, in preference to general RNA extraction and DNA extraction protocols.

Caution: See Cautions Appendix for appropriate handling of materials marked with ▼.

Isolation of Total DNA

From blood

Acid Citrate Dextrose (ACD), Solution B (Storage) VI MC3

0.48% w/v citric acid
1.32% w/v sodium citrate
1.47% w/v glucose
For freshly drawn or frozen blood samples

RBC (Red Blood Cell) Lysis Buffer MC3

20 mM Tris-Cl (pH 7.6)
Store the buffer at room temperature.

Cell Lysis Buffer   MC3

10 mM Tris-Cl (pH 8.0)
1 mM EDTA (pH 8.0)
0.1% (w/v) SDS ▼

Store the buffer at room temperature, but chill an aliquot to 0°C.

Potassium Acetate Solution   MC3

60 ml of 5 M potassium acetate
11.5 ml of glacial acetic acid ▼
28.5 ml of H_2O

The resulting solution is 3 M with respect to potassium and 5 M with respect to acetate. Store the buffer at room temperature.

Purification of High-molecular-weight DNA

Dialysis Buffer I   MC3

20 mM Tris-Cl (pH 8.0)
0.1 M NaCl
10 mM EDTA (pH 8.0)

Prepare 6 liters of dialysis buffer 1. Store at 4°C.

Dialysis Buffer 2   MC3

10 mM Tris-Cl (pH 8.0)
10 mM NaCl
0.5 mM EDTA (pH 8.0)

Prepare 6 liters of dialysis buffer 2. Store at 4°C.

Formamide Denaturation Buffer   MC3

20 mM Tris-Cl (pH 8.0)
0.8 M NaCl
80% (v/v) formamide ▼

Many batches of reagent-grade formamide are sufficiently pure to be used without further treatment. However, if any yellow color is present, the formamide should be deionized.

By Spooling

Cell Lysis Solution MC3

6 M guanidinium hydrochloride ▼
0.1 M sodium acetate (pH 5.5) ▼

Cell Lysis Buffer MC3

10 mM Tris-Cl (pH 7.5)
10 mM NaCl
10 mM EDTA (pH 8.0)
0.5% (w/v) Sarkosyl

Add proteinase K to the lysis buffer to a final concentration of 1 mg/ml just before use.

Sarkosyl, an anionic detergent, is usually supplied by the manufacturer as a 30% solution in H_2O. It is less prone than SDS to precipitate from solutions of high ionic strength. However, it is also a less effective detergent than SDS.

NaCl/Ethanol Solution MC3

Add 150 μl of 5 M NaCl per 10 ml of absolute ethanol. Store the NaCl/ethanol solution at −20°C.

PCR (Polymerase Chain Reaction) Lysis Solution A MC3

67 mM Tris-Cl (pH 8.8)
16.6 mM ammonium sulfate ▼
5 mM β-mercaptoethanol ▼
6.7 mM $MgCl_2$ ▼
6.7 μM EDTA (pH 8.0)
1.7 μM SDS ▼
50 μg/ml proteinase K

PCR Lysis Solution B MC3

10 mM Tris-Cl (pH 8.3)
50 mM KCl ▼
2 mM $MgCl_2$ ▼
0.45% (v/v) Nonidet P-40
0.45% (v/v) Tween-20
20 μg/ml proteinase K

Lysate Buffer MC3

670 mM Tris-Cl (pH 8.8)
166 mM ammonium sulfate ▼
1 mg/ml bovine serum albumin

SNET MC3

20 mM Tris-Cl (pH 8.0)
5 mM EDTA (pH 8.0)
400 mM NaCl
1% (w/v) SDS ▼

Sterilize the solution by filtration through a 0.45-μm nitrocellulose filter. Store the sterile solution in 50-ml aliquots at room temperature.

SNET Lysis Buffer Volumes MC3

Age of mouse	Amount of tissue	Type of tube	Volume of SNET lysis buffer (ml)
Newborn	entire tail (1 cm)	microfuge	0.5
10 days old	distal one-third	microfuge	0.5
Weanling (3–4 weeks)	6–10 mm	17 × 100-mm polypropylene	4.0
Any age	100 mg of fresh tissue	17 × 100-mm polypropylene	4.0

From Mouse

Lysis Buffer for a 96-well Plate MME3

10 mM Tris (pH 7.5) ▼
10 mM EDTA
10 mM NaCl
0.5% (w/v) Sarcosyl

Add 1 mg/ml proteinase K just before use.

Lysis Buffer for a 24-well Plate   MME3

100 mM Tris-HCl (pH 8.5)
5 mM EDTA
0.2% SDS ▼
200 mM NaCl
100 µg/ml proteinase K (20 mg/ml proteinase K stock is stored at −20° and added to the lysis buffer before use)

From Yeast

STES Buffer   MC3

0.2 M Tris-Cl (pH 7.6)
0.5 M NaCl
0.1 % (w/v) SDS ▼
0.01 M EDTA
Store at room temperature.

From Plant Cells

CTAB DNA Extraction Buffer   ARA

100 mM Tris-Cl (pH 8.0)
20 mM EDTA (pH 8.0)
1.4 M NaCl
2% (w/v) CTAB (cetyltrimethyl ammonium bromide) ▼
1% PVP 40,000 (polyvinyl pyrrolidone) ▼
After autoclaving, store at room temperature.

EB Extraction Buffer   ARA

100 mM Tris (pH 8.0) ▼
50 mM EDTA (pH 8.0)
500 mM NaCl
10 mM β-mercaptoethanol (add just before use 7.2 µl/10 ml) ▼

BTE   ARA

50 mM Tris (pH 8.0) ▼
1 mM EDTA (pH 8.0)

Quick DNA Prep Extraction Buffer   ARA

200 mM Tris-Cl (pH 7.5)
250 mM NaCl
25 mM EDTA
0.5% SDS ▼

Isolation of Total RNA

Before performing all labeling, first-strand synthesis, and amplification protocols using RNA, it is advisable to check the quality of RNA using standard electrophoretic protocols (e.g., Agilent Bioanalyser). All buffers and reagents used for RNA analysis and preparation should first be treated with DEPC. Information on storage and handling of RNA may be found in the first volume of *Lab Ref*.

From Mammalian Cells

Solution D (Denaturing Solution)   MC3

4 M guanidinium thiocyanate ▼
25 mM sodium citrate · 2H$_2$O ▼
0.5% (w/v) sodium lauryl sarcosinate ▼
0.1 M β-mercaptoethanol ▼

Dissolve 250 g of guanidinium thiocyanate in 293 ml of H$_2$O, 17.6 ml of 0.75 M sodium citrate (pH 7.0), and 26.4 ml of 10% (w/v) sodium lauryl sarcosinate. Add a magnetic bar and stir the solution on a combination heater-stirrer at 65°C until all ingredients are dissolved. Store solution D at room temperature and add 0.36 ml of 14.4 M stock β-mercaptoethanol per 50 ml of solution D just before use. Solution D may be stored for months at room temperature but it is sensitive to light. Note that guanidinium will precipitate at low temperatures.

Amounts of Solution D Required to Extract RNA from Cells and Tissues MC3

Amount of tissue or cells	Amount of solution D (ml)
100 mg of tissue	3
T-75 flask of cells	3
60-mm plate of cells	1
90-mm plate of cells	2

From Yeast

Acid Phenol MICRO

Acid phenol (pH 5.0–5.1) ▼

To make acid phenol, heat a fresh bottle (100 g) of crystalline phenol to 65°C. The liquefied phenol should be clear and free of color. Discard batches of phenol that are yellow or pink. Add 20 ml of a solution of 10 mM sodium acetate (pH 5.1), 50 mM NaCl, and 1.0 mM EDTA (pH 8.0). Mix the contents of the bottle by stirring. Let the liquefied phenol cool to room temperature and then add another 10 ml of the buffer. Wrap the bottle in aluminum foil and store it at 4°C.

AE Buffer MICRO

50 mM sodium acetate (pH 6.0)
10 mM EDTA

From *E. coli*

Ethanol/Phenol Stop Solution MICRO

H_2O-saturated phenol (pH <7.0) in ethanol (5% v/v).

The recipe for this solution was provided by Jon Berstein (University of California, San Francisco).

Water-saturated Phenol (pH <7.0) MICRO

Prepare the phenol solution from Ultra-Pure phenol (redistilled, crystalline; Invitrogen/GIBCO 15509). Note that H_2O-saturated acid phenol should be used for RNA extractions.

Lysozyme Solution   MICRO

500 µg/ml in 10 mM Tris (pH 8.0) ▼
1 mM EDTA (pH 8.0)

Prepare the lysozyme solution fresh just before use.

5x DNase I Buffer   MICRO

50 mM $MgCl_2$ ▼
50 mM Tris-Cl (pH 7.5)
5 mM EDTA (pH 8.0)
5 mM dithiothreitol (DTT) ▼

From Plant Tissue

RNA Precipitation Solution   MICRO

0.8 M sodium citrate ▼
1.2 M NaCl

No adjustment of pH is required.

Monophasic Lysis Reagents   MC3

Reagent	Commercial supplier
Trizol reagent	Life Technologies
TRI reagent	Molecular Research Center
Isogen	Nippon Gene, Toyama, Japan
RNA-Stat-60	Tel-Test

When using commercial reagents for the simultaneous isolation of RNA, DNA, and protein, we recommend following the manufacturer's instructions. In most cases, these differ little from the generic instructions.

Extraction Buffer   MICRO

2% (w/v) CTAB ▼
2% (w/v) PVP K 30 ▼
100 mM Tris-HCl (pH 8.0)
25 mM EDTA
2.0 M NaCl
0.5 g/liter spermidine
2% β-mercaptoethanol ▼

Mix all ingredients, except β-mercaptoethanol. Divide the solution into small aliquots and sterilize by autoclaving on liquid cycle for 15 minutes. Add β-mercaptoethanol to the extraction buffer just before use.

Lithium Chloride   MICRO

LiCl (8 M, F.W. = 42.39) ▼

Add 33.9 g of LiCl to 50 ml of H_2O. Adjust the volume of the solution to 100 ml. After all of the LiCl has dissolved, sterilize the solution by autoclaving for 15 minutes on liquid cycle.

SSTE   MICRO

1.0 M NaCl
0.5% (w/v) SDS ▼
10 mM Tris-HCl (pH 8.0)
1 mM EDTA (pH 8.0)

From Fixed Tissues

DNA Extraction Buffer   MICRO

50 mM Tris-Cl (pH 7.6)
1 mM EDTA (pH 8.0)
1% Tween-20
2 mg/ml proteinase K

Using Ultracentrifugation

CsCl₂ Solution   MICRO

CsCl (5.7 M)/0.01 M EDTA (pH 7.5)

The CsCl/EDTA solution is made in 100-ml batches by dissolving 96.0 g of CsCl▼ in 90 ml of 0.01 M EDTA (pH 7.5) and adding DEPC to a final concentration of 0.1%. Allow the solution to stand for 30 minutes and then autoclave for 20 minutes at 15 psi (1.05 kg/cm²). When the solution has cooled, adjust the volume to 100 ml with DEPC-treated H₂O.

Guanidinium Thiocyanate Homogenization Buffer   MICRO

4 M guanidinium thiocyanate (M_r = 118.1) ▼
0.1 M Tris-Cl (pH 7.5)
1% β-mercaptoethanol ▼

Dissolve 50 g of guanidinium thiocyanate in 10 ml of 1 M Tris-Cl (pH 7.5) and add H₂O to 100 ml. Filter the solution through a Whatman No. 1 filter or equivalent. The solution is stable and can be stored indefinitely at room temperature. Just before use, add β-mercaptoethanol to a final concentration of 1% (0.14 M).

Selection of Poly(A) RNA from Total

By Oligo dT

2x Column-loading Buffer   MC3

40 mM Tris-Cl (pH 7.6)
1 M NaCl
2 mM EDTA (pH 8.0)
0.2% (w/v) sodium lauryl sarcosinate ▼

Prepare as described below.

Make up Tris-Cl (pH 7.6) from a fresh bottle in autoclaved, DEPC-treated H₂O. Prepare NaCl and EDTA in 0.1% DEPC▼ in H₂O. Store for at least 12 hours at 37°C and autoclave the mixture for 15 minutes at 15 psi (1.05 kg/cm²) on liquid cycle. To prepare sterile column-loading buffer, mix appropriate amounts of RNase-free stock solutions of Tris-Cl (pH 7.6), NaCl, and EDTA (pH 8.0) in an RNase-free container. Allow the solution to cool to approximately 65°C, and then add sodium lauryl sarcosinate from a 10% stock solution that has been heated for 30 minutes to 65°C.

Alternatively, substitute 0.05 M sodium citrate for Tris-Cl, and treat the sodium citrate/NaCl/EDTA mixture and sodium lauryl sarcosinate with DEPC. Store column-loading buffer at room temperature.

Elution Buffer MC3

10 mM Tris-Cl (pH 7.6)
1 mM EDTA (pH 8.0)
0.05% SDS ▼

The stock solutions of Tris-Cl and EDTA used to make elution buffer should be freshly autoclaved (15 minutes at 15 psi [1.05 kg/cm²] on liquid cycle) and then diluted with the appropriate amount of sterile DEPC-treated H_2O. Add the SDS from a concentrated stock solution (10% or 20%) made in sterile DEPC-treated H_2O.

IMPORTANT: Do not attempt to sterilize elution buffer by autoclaving because it froths excessively.

By Batch Chromatography

Absorption/Washing Buffer MC3

This buffer is TES containing 0.5 M NaCl.

Na-TES MICRO

This is TES containing 0.5 M NaCl.

TES MICRO

10 mM Tris-HCl (pH 7.5)
1 mM EDTA (pH 8.0)
SDS (0.1% w/v) ▼

Selection of Polysomal RNA

Cyclohexamide MICRO

Cycloheximide (10 mg/ml in H_2O) (Sigma-Aldrich) ▼

Cycloheximide is unstable in alkaline solutions. Cycloheximide inhibits protein synthesis in eukaryotes, but not prokaryotes, by interacting directly with the translocase enzyme and blocking translocation of peptidyl tRNA from the A site to the P site within the ribosome. The drug is a competitive inhibitor of FKBP and related enzymes and an inducer of apoptosis.

Gradient Buffer   MICRO

150 mM KCl▼
5 mM MgCl$_2$ ▼
50 mM Tris-Cl (pH 7.4)

Autoclave the solution before use.

Hypotonic Lysis Buffer   MICRO

10 mM KCl▼
1.5 mM MgCl$_2$ ▼
10 mM Tris-Cl (pH 7.4)

Sterilize the solution by autoclaving. Add cycloheximide to a final concentration of 10 µg/ml just before use.

RNA Precipitation Solution   MICRO

0.8 M sodium citrate ▼
1.2 M NaCl

Sucrose Buffer   MICRO

Prepare these solutions by appropriate dilution of the 2.5 M sucrose stock with gradient buffer. Each gradient requires 13 ml of 1.95 M sucrose buffer and 6 ml of 1.3 M sucrose buffer. Just before use in a gradient, add cycloheximide ▼ to a final concentration of 10 µg/ml.

It is a good idea to test these solutions by constructing a mock gradient and checking for proper layering before constructing any gradients containing actual lysate.

Sucrose Gradient Buffer   MICRO

This solution is extremely viscous and must be made carefully.

1. Add 855.75 g of sucrose in approximately 150-g increments to 250 ml of gradient buffer heated to 65°C in a 2-liter beaker.

2. Stir the buffer on a hot plate using an appropriately sized stirring bar and maintain the temperature of the solution near 65°C. If the solution becomes too viscous, add a small amount of gradient buffer (65°C).

3. Once all of the sucrose has been added, adjust the final volume to 1 liter with gradient buffer (65°C). It is possible, although not usually necessary, to measure the concentration of sucrose in the solution using a refractometer and to adjust the strength of the solution accordingly.

4. Just before use in a gradient, add cycloheximide to a final concentration of 10 µg/ml. Each gradient requires 15 ml of the 2.5 M sucrose buffer.

Isolation and Amplification of RNA from Single Cells and Small Amounts of Tissue

The buffers detailed in this section refer to specific Cold Spring Harbor Laboratory Press protocols. Commercial products containing some or all reagents are available (e.g., Ambion Message Amp II). Before all amplification protocols, it is advisable to check the quality of RNA either before or after amplification, using any available electrophoretic-based method (e.g., Agilent Bioanalyser).

10x RNA Amplification Buffer MICRO

400 mM Tris-base (pH 7.5)
70 mM $MgCl_2$ ▼
100 mM NaCl
20 mM spermidine

Filter before adding spermidine.

10x Second-strand Buffer MICRO

1 M Tris-base (pH 7.4)
200 mM KCl ▼
100 mM $MgCl_2$ ▼
400 mM $(NH_4)_2SO_4$ ▼

Store in small aliquots at –20°C.

10x AMV-RT MICRO

500 mM Tris-base (pH 8.3)
1.2 M KCl ▼
100 mM $MgCl_2$ ▼

10x KFI MICRO

200 mM Tris-base (pH 7.5)
100 mM $MgCl_2$ ▼
50 mM NaCl
50 mM DTT ▼

DTT may be omitted from the buffer and added directly to the reaction because it tends to precipitate out of the buffers over time.

10x Nuclease S1 Buffer MICRO

2 M NaCl
500 mM sodium acetate (pH 4.5) ▼
10 mM ZnSO₄ ▼

Hybridization Analysis

This section contains buffers used in the fixation and hybridization of nucleic acids following electrophoretic separation.

10x BPTE Electrophoresis Buffer MC3

100 mM PIPES (piperazine-1,4-bis[2-ethanesulfonic acid])
300 mM Bis-Tris (bis[2-hyroxyethyl]iminotris[hydroxymethyl]methane)
10 mM EDTA (pH 8.0)

The final pH of the 10x buffer is approximately 6.5.

Prepare the 10x buffer by adding 3 g of PIPES (free acid), 6 g of Bis-Tris (free base), and 2 ml of 0.5 M EDTA (pH 8.0) to 90 ml of distilled H₂O, then treating the solution with DEPC ▼ , final concentration 0.1%, for 1 hour at 37°C, and then autoclaving.

10x Formaldehyde Gel-loading Buffer MC3

50% glycerol (diluted in DEPC-treated H₂O)
10 mM EDTA (pH 8.0)
0.25% (w/v) bromophenol blue ▼
0.25% (w/v) xylene cyanol FF ▼

10x MOPS Buffer MC3

0.2 M MOPS (pH 7.0) ▼
20 mM sodium acetate ▼
10 mM EDTA (pH 8.0)

Dissolve 41.8 g of MOPS (3-[N-morpholino]propanesulfonic acid) in 700 ml of sterile DEPC-treated H₂O. Adjust the pH to 7.0 with 2 N NaOH. Add 20 ml of DEPC-treated 1 M sodium acetate and 20 ml of DEPC-treated 0.5 M EDTA (pH 8.0). Adjust the volume of the solution to 1 liter with DEPC-treated H₂O. Sterilize the solution by filtration through a 0.45-μm Millipore filter, and store it at room temperature protected from light. The buffer yellows with age if it is exposed to light or is autoclaved. Straw-colored buffer works well, but darker buffer does not.

Soaking Solution MC3

For charged membranes, use 0.01 N NaOH ▼ combined with 3 M NaCl; for uncharged membranes, use 0.05 N NaOH.

Transfer Buffer MC3

For alkaline transfers to charged membranes, use 0.01 N NaOH ▼ with 3 M NaCl; for neutral transfers to uncharged membranes, use 20x SSC.

Prehybridization Solution MC3

0.5 M sodium phosphate (pH 7.2) ▼
7% (w/v) SDS ▼
1 mM EDTA (pH 7.0)

Background is often a problem when using nylon membranes. Of the large number of hybridization buffers that have been described in the literature, those containing high concentrations of SDS are the most effective at suppressing background while preserving high sensitivity.

Dot and Slot Blots

RNA Denaturation Solution MC3

660 μl of formamide ▼
210 μl of 37% (w/v) formaldehyde ▼
130 μl of 10x MOPS electrophoresis buffer (pH 7.0) ▼

Formaldehyde is supplied as a 37–40% w/v (12.3 M) solution that may contain a stabilizer such as methanol (10–15%). Formaldehyde oxidizes readily to formic acid when exposed to air. If the pH of the formaldehyde solution is acidic (<pH 4.0) or if the solution is yellow, the stock solution should be deionized by treatment with a mixed bed resin, such as Bio-Rad AG-501-X8 or Dowex XG8 before use.

Purchase a distilled deionized preparation of formamide reagent and store in small aliquots under nitrogen at –20°C.

Tissue Arrays

Optimizing DIG-ISH   MICRO

Parameter	Reason to vary	Suggested range of optimizing[a]
PREHYBRIDIZATION Unmasking mRNA by digestion with proteinase K	different tissue; different fixation method; insufficient probe staining	1, 5, and 10 µg/ml (1 µg/ml)
Probe concentration	different/new probe	100 and 200 ng/ml (200 ng/ml) 25–100 ng/ml (amplified ISH) (50 ng/ml, amplified ISH)
Hybridization temperature	different/new probe; different tissue; different fixation method	42°C, 50°C, and 60°C
POSTHYBRIDIZATION RNase treatment	only when excessively high background cannot be eliminated	RNase treatment at 20 µg/ml (not performed)
Anti-DIG-AP antibody incubation	insufficient probe signal; excessive background staining	1:500–1:1500 (1:1000)
Anti-DIG-HRP (amplified ISH)	optimizing amplification parameters	1:50–1:500 (1:100)
Anti-DNP-AP (amplified ISH)	optimizing amplification parameters	1:25–1:500 (1:50)
DNP amplification substrate	optimizing amplification parameters	5–30 minutes (10 min)

[a]Parentheses indicate suggested starting point.

Maleic Acid Buffer   MICRO

100 mM maleic acid ▼
150 mM NaCl

NBT/BCIP Buffer (Nitroblue Tetrazolium/5-bromo-4-chloro-3-indolyl Phosphate ▼)   MICRO

100 mM Tris (pH 9.5) ▼
50 mM MgCl$_2$ ▼
100 mM NaCl

NBT/BCIP Color Substrate ▼   MICRO

10 μl of NBT/BCIP buffer ▼
200 μl of NBT/BCIP solution

Prehybridization/Hybridization Buffer   MICRO

formamide (deionized) ▼	23.8 ml
1 M Tris (pH 7.4) ▼	0.95 ml
250 mM EDTA (pH 8)	0.19 ml
5 M NaCl	3.00 ml
dextran sulfate	4.80 g
50x Denhardt's solution	0.95 ml
H$_2$O (DEPC-treated, sterile, distilled)	6.35 ml
Total volume	40.0 ml

This buffer is best prepared just before use. Alternatively, the buffer can be made and stored in small volumes for up to 3 months at –20°C with satisfactory results. Do not refreeze the buffer once it has thawed.

Ribonucleic Acid Solution   MICRO

herring sperm DNA (10 mg/ml)	1 ml
tRNA (25 mg/ml)	1 ml
yeast total RNA (25 mg/ml)	1.25 ml
H$_2$O (DEPC-treated, sterile, distilled)	748 μl
Total volume	4 ml

Prepare solution in 500-μl aliquots.

10x Proteinase K Stop Solution MICRO

2% glycine in PBS

1x Proteinase K Digest Buffer MICRO

100 mM Tris (pH 8) ▼
50 mM EDTA (pH 8)

Make up to 1 liter and autoclave.

DNA Microarrays

Cell Lines Used in Stanford "Common RNA Reference" MICRO

Name	Description	Growth properties	ATCC catalog no.
MCF7	breast adenocarcinoma-derived cell line	adherent	ATCC HTB-22
Hs578T	breast adenocarcinoma-derived cell line (stromal-like)	adherent	ATCC HTB-126
NTERA2	teratoma-derived cell line	adherent	ATCC CRL-1973
Colo205	colon tumor-derived cell line	mixed	ATCC CCL-222
OVCAR-3	ovarian tumor-derived cell line	adherent	ATCC HTB-161
UACC-62	melanoma-derived cell line	adherent	
MOLT-4	T cell leukemia-derived cell line	suspension	ATCC CRL-1582
RPMI 8226	multiple myeloma-derived cell line	suspension	ATCC CCL-155
NB4+ATRA	acute promyelocytic leukemia-derived cell line	suspension	
SW872	liposarcoma-derived cell line	adherent	ATCC HTB-92
HepG2	liver tumor-derived cell line	adherent	ATCC HB-8065

First-strand cDNA Buffer MICRO

First-strand cDNA buffer (as supplied by the manufacturer of the enzyme)

If buffer supplied by the manufacturer is not available, use the following 5x first-strand buffer:

250 mM Tris-Cl (pH 8.3)
375 mM KCl ▼
15 mM $MgCl_2$ ▼

Preparation of Nucleic Acid (Target) for Array Hybridization Analysis

Labeling of Genomic DNA

2.5x Random Primer Mix   MICRO

125 mM Tris-Cl (pH 6.8)
12.5 mM $MgCl_2$ ▼
25 mM β-mercaptoethanol ▼
750 µg/ml random octamers

These components can be assembled separately, or the complete buffer can be purchased as part of the BioPrime DNA Labeling System (Invitrogen 18094-011).

10x One-Phor-All Buffer Plus   MICRO

100 mM Tris-acetate (pH 7.5)
100 mM magnesium acetate ▼
500 mM potassium acetate

Store the buffer at 4°C.

One-Phor-All Buffer Plus is a universal buffer available from GE Healthcare 27-0901-02.

Preparation of Fluorescently Labeled Nucleic Acid

Synthesis of First-strand cDNA   MICRO

Add to each of the annealing mixtures:

5x first-strand buffer (SUPERSCRIPT reverse transcriptase kit)	8 µl
10x dNTP/low dTTP solution	4 µl
Cy5- or Cy3-labeled dUTP (1 mM)	4 µl
0.1 M DTT ▼	4 µl
RNasin (30 units/µl)	1 µl
SUPERSCRIPT II (200 units/µl)	2 µl

Reverse transcriptases are very sensitive to denaturation at air/liquid interfaces, so be very careful to avoid foaming when mixing the components of the reaction. This is sometimes difficult because many preparations of reverse transcriptases are supplied in buffers containing nonionic detergents such as Nonidet P-40 or Triton X-100.

Mo-MLV (SUPERSCRIPT II) is temperature sensitive and should be stored at −20°C until the last possible moment. Return to the freezer at the first possible opportunity.

Micron YM-column Usage MICRO

Microcon YM-100 (Millipore 42412) or Microcon YM-30 (Millipore 42411)
YM-100 centrifugal devices retain >90% of single-stranded cDNAs whose
sizes exceed 300 nucleotides. For YM-30 devices, the nominal cutoff for single-stranded cDNAs is about 50 nucleotides.

Property	Microcon YM-100	Microcon YM-30
Color of device	blue	clear
Filtration area (cm^2)	0.32	0.32
Nominal cutoff (daltons)	100,000	30,000
Maximum allowable *g* force	500*g* in a fixed-angle rotor	14,000*g* in a micro-fuge
Time of centrifugation	5–15 min at 25°C	5–8 min at 25°C
Capacity of device (ml)	0.5	0.5
Retention volume (µl)	10	10

Amplification of mRNA

10x RNA Amplification Buffer MICRO

400 mM Tris-Cl (pH 7.5)
70 mM MgCl$_2$ ▼
100 mM NaCl
20 mM spermidine
Filter before adding spermidine.

10x Second-strand Buffer MICRO

1 M Tris-Cl (pH 7.4)
200 mM KCl ▼
100 mM MgCl$_2$ ▼
400 mM (NH$_4$)$_2$SO$_4$ ▼

Labeling mRNA with Biotin Nucleotides

5x Fragmentation Buffer MICRO

Reagent	Amount	Final concentration (mM)
1 M Tris-acetate (pH 8.1)	4.0 ml	200
MgOAc	0.64 g	150
KOAc	0.98 g	500

Recovery of Unincorporated Dyes by High-performance Liquid Chromatography (HPLC)

High-salt Buffer (HSB) MICRO

1 M triethylammonium acetate (pH 8.0)
10% acetonitrile ▼

Pass the buffer through a nylon 0.22-μm filter. Sparge the buffer in helium immediately before use in HPLC.

Low-salt Buffer (LSB) MICRO

25 mM triethylammonium acetate (pH 8.0)
10% acetonitrile ▼

Prepare a 40-fold dilution of HSB and add acetonitrile to 10%. Pass the buffer through a nylon 0.22-μm filter. Sparge the buffer in helium immediately before use in HPLC.

Alexa Dyes

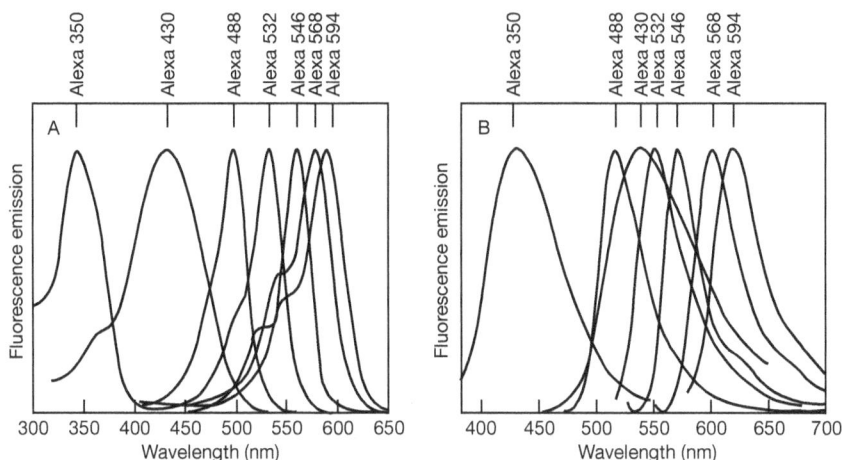

MICRO Normalized absorption (*A*) and emission (*B*) spectra of the Alexa dyes conjugated to goat antimouse antibody. With the exception of a few nanometers shift of the excitation/emission maxima toward the red region of the spectrum, the profile of the conjugates' spectra closely resembles that of the free Alexa dyes. (Redrawn, with permission, from N. Panchuk-Voloshina et al. 1999. Alexa dyes, a series of new fluorescent dyes that yield exceptionally bright, photostable conjugates. *J. Histochem. Cytochem.* 47: 1179–1188.)

Preparation of Nucleic Acid (Probe) for Array Hybridization Analysis

Amplification of Probes from Bacterial Clone Sets

Primers for Amplification   MICRO

Vector	Libraries used	Forward and reverse primers
pT3-T7Pac	Soares	Fwd: ctgcaaggcgattaagttgggtaac Rev: gtgagcggataacaatttcacacaggaaaacagc
pCMV-Sport 6	NCI-GAP	Fwd: ctgcaaggcgattaagttgggtaac Rev: gtgagcggataacaatttcacacaggaaaacagc
pBluescriptSK-	NCI-GAP; Stratagene	Fwd: ctgcaaggcgattaagttgggtaac Rev: gtgagcggataacaatttcacacaggaaaacagc
LafmBA	Soares 1 NIB	Fwd: ctgcaaggcgattaagttgggtaac Rev: gtgagcggataacaatttcacacaggaaaacagc
pCMV-Sport	Beddington embryonic region	Fwd: ggagagagctatgacgtcgc Rev: gacactatagaaggtacgcctgc
pCMV-Sport 1	NIA	Fwd: ccagtcacgacgttgtaaaacgac Rev: gtgtggaattgtgagcggataacaa
pCMV-Sport 2	Life Tech embryo	Fwd: ggagagagctatgacgtcgc Rev: gacactatagaaggtacgcctgc
pOT2A	BDGP *Drosophila* library	Fwd: aatgcaggttaacctggcttatcg Rev: aacgcggctacaattaatacataacc
Vectors with M13 priming sites	wide range	Fwd: gttttcccagtcacgacgttg Rev: tgagcggataacaatttcacacag

Isolation and Amplification of Probes from Yeast DNA Microarrays

Lysis Buffer (Prepare 100 ml)   MICRO

0.9 M sorbitol
0.1 M Tris-Cl (pH 8.0)
0.1 M EDTA

Resuspension Buffer 1 (Prepare 50 ml)   MICRO

50 mM Tris-Cl (pH 8.0)
20 mM EDTA

Resuspension Buffer 2   MICRO

50 mM Tris-Cl (pH 8.0)
1 mM EDTA

10x Amplification Buffer MICRO

500 mM KCl ▼
100 mM Tris-HCl (pH 8.3)
MgCl₂ ▼ is added separately. Store the buffer in small aliquots at –20°C.

Guide for Fixing Slides MICRO

To fix by baking: Place the slides in an oven preset to 80°C for an appropriate period of time.

Baking times vary from 1 to 4 hours. Longer times are required when baking alone, rather than in combination with UV irradiation, which is used to promote cross-linking.

To fix by UV cross-linking: Irradiate the slides by exposure to 60–200 mJ as appropriate.

For slide coatings that require cross-linking to achieve covalent attachment of DNA to the substrate (e.g., for poly-L-lysine-coated slides), it is critical that cross-linking be performed effectively. Because the degree of cross-linking may affect the strength of the signal, it is worth experimenting to find the optimal time of baking and/or exposure to UV irradiation when using new types of substrates. Make sure that the UV lamps have not deteriorated over time. Failure to adequately fix DNA to the slides can be a common cause of "comets."

(*Optional*) Immediately snap-dry the slide by heating it for 20–30 seconds (5 sec for poly-L-lysine-coated slides) on a hot plate set at 100–140°C.

This treatment is usually recommended for amino-silane, CMT-GAPS II, and poly-L-lysine-coated slides. A 20–30-second snap-heat step after UV cross-linking usually eliminates the comet-tail problem with CMT-GAPS II and with amino-silane-coated slides. The slides should then be air cooled to avoid shattering and rinsed in a large volume of distilled H₂O.

Recommended Conditions for Fixing DNA to the Slide Substrate

Type of slide/slide coating	Baking	UV cross-linking
Poly-L-lysine	X	60 mJ
Telechem SuperAmine	1 hour	optional
Telechem SuperAldehyde	X	X
Corning MicroTechnology	2–4 hours	200 mJ
Gamma amino propyl silane (CMT-GAPS II)		

This table provides examples of recommended handling conditions for some commonly used substrates. X signifies that the method is inappropriate, not needed, or not the preferred method for the specified slide coating.

Other Quantitative Methods

In recent years, quantitative analysis of gene expression has been dominated by protocols in array hybridization and quantitative PCR; however, S1 RNA protection assays also provide accurate quantitative measures of RNA abundance.

Nuclease S1 RNA Protection

10x Annealing Buffer MC3

100 mM Tris-Cl (pH 7.5)
100 mM MgCl$_2$ ▼
0.5 M NaCl
100 mM dithiothreitol

Gel Elution Buffer MC3

0.5 M ammonium acetate ▼
1 mM EDTA (pH 8.0)
0.1% (w/v) SDS ▼

Hybridization Buffer MC3

40 mM PIPES (pH 6.4)
0.1 mM EDTA (pH 8.0)
0.4 M NaCl

Use the disodium salt of PIPES (piperazine-N,N'-bis[2-ethanesulfonic acid]) to prepare the buffer and adjust the pH to 6.4 with 1 N HCl.

Nuclease S1 Stop Mixture MC3

4 M ammonium acetate ▼
50 mM EDTA (pH 8.0)
50 µg/ml carrier RNA

Store the nuclease S1 stop mixture in aliquots at –20°C.

Nuclease S1 Digestion Buffer MC3

0.28 M NaCl
0.05 M sodium acetate (pH 4.5) ▼
4.5 mM ZnSO$_4$·7H$_2$O ▼

Store aliquots of nuclease S1 buffer at –20°C and add nuclease S1 to a concentration of 500 units/ml just before use.

Percentage of Gel for Purifying Various DNA Fragments MC3

% Polyacrylamide/urea gel	Size of band (nt)
4	>250
6	60–250
8	40–120
10	20–60
12	10–50

Expected Mobilities of Tracking Dyes MC3

% Polyacrylamide/urea gel	Xylene cyanol (nt)	Bromophenol blue (nt)
4	155	30
6	110	25
8	75	20
10	55	10

Tracking dyes can serve as useful size standards on denaturing polyacrylamide gels. The table indicates the approximate sizes of tracking dyes (in nucleotides) on gels of different polyacrylamide concentrations.

Hybridization Buffer MC3

40 mM PIPES (pH 6.8)
1 mM EDTA (pH 8.0)
0.4 M NaCl
80% deionized formamide ▼

Use the disodium salt of PIPES (piperazine-*N*,*N*′-bis[2-ethanesulfonic acid]) and adjust the pH with 1 N HCl.

RNA Gel-loading Buffer MC3

95% deionized formamide ▼
0.025% (w/v) bromophenol blue ▼
0.025% (w/v) xylene cyanol FF ▼
5 mM EDTA (pH 8.0)
0.025% (w/v) SDS ▼

Purchase a distilled deionized preparation of formamide and store in small aliquots under nitrogen at –20°C.

10x Transcription Buffer   MC3

0.4 M Tris-Cl (pH 7.5)
0.1 M NaCl
60 mM MgCl₂ ▼
20 mM spermidine

Store in aliquots at –20°C.

Carrier RNA (1 mg/ml)   MC3

Prepare a stock of carrier RNA (1 mg/ml) by dissolving commercially available yeast tRNA at a concentration of 10 mg/ml in sterile TE (pH 7.6) containing 0.1 M NaCl. Extract the solution twice with phenol (equilibrated in Tris-Cl [pH 7.6]) and twice with chloroform ▼ . Recover RNA by precipitation with 2.5 volumes of ethanol at room temperature. Dissolve the precipitated RNA at a concentration of 10 mg/ml in sterile TE (pH 7.6), divide the stock into small aliquots, and store them at –20°C.

NOTES

Section 2
RNAi

RNAi

This section contains buffers that are used in RNA silencing and interference. Topics not covered include commonly used commercial buffers (e.g., Ambion mirVana) in the processing and isolation of endogenous microRNAs.

Caution: See Cautions Appendix for appropriate handling of materials marked with ▼.

RNAi in *C. elegans*

The following buffers deal with injection-based protocols, although feeding *C. elegans* with RNase-*E. coli* expressing double-stranded RNA may also be effective.

10x M9 RNAi

30 g of KH_2PO_4▼
60 g of K_2HPO_4▼
50 g of NaCl
After autoclaving, add 10 ml of 1 M $MgSO_4$▼.

NGM (Worm Culturing) RNAi

3 g of NaCl
17 g of agar
2.5 g of peptone
1 ml of 5 mg/ml cholesterol
After autoclaving, add:
1 ml of 1 M $CaCl_2$▼
1 ml of 1 M $MgSO_4$▼
25 ml of 1 M K_2HPO_4/KH_2PO_4 (pH 6)▼
As a food source for the nematodes, seed plates with the *E. coli* strain OP50.

29

Hybridization Buffer (RNase Protection) RNAi

50 mM PIPES (pH 6.4)
1.3 mM EDTA
0.5 M NaCl
67% formamide ▼

Low-stringency Hybridization Buffer (Northern Blotting) RNAi

0.36 M Na_2HPO_4 ▼
0.14 M NaH_2PO_4 ▼
1 mM EDTA
7% SDS ▼

The pH of the entire mix should be 7.2. Do not autoclave.

RNAi in *D. melanogaster*

These buffers are used for the preparation of *Drosophila* embryos and for diagnostic evaluation of dsRNA expression.

Hypotonic Buffer RNAi

1 mM $MgCl_2$ ▼
300 mM potassium acetate

Lysis Buffer (for Drosophila *Embryo Extracts*) RNAi

5 mM dithiothreitol ▼
30 mM HEPES-KOH (pH 7.4) ▼
2 mM magnesium acetate ▼
1 mg/ml Pefabloc SC (Roche)
100 mM potassium acetate

Lysis Buffer (for siRNA-primed RNA synthesis) RNAi

100 mM potassium acetate
30 mM HEPES-KOH (pH 7.4) ▼
2 mM magnesium acetate ▼
5 mM DTT ▼
1 mg/ml Pefabloc SC (Roche)

2x PK Buffer RNAi

200 mM Tris-HCl (pH 7.5)
25 mM EDTA
300 mM NaCl
SDS (2% w/v) ▼

RNAi in Plants

In this case, infiltration medium is used to facilitate the *Agrobacterium* transfer of dsRNA producing constructs into plants.

Infiltration Medium RNAi

10 mM 2-(*N*-morpholino) ethanesulfonic acid (MES) ▼
10 mM MgCl$_2$ ▼
150 mM Acetosyringone (Apollo Scientific)

RNAi in Mammalians

The following are used in the preparation of dsRNA for introduction into mammalian cells, as well as investigation of protein knockdown (by western blotting).

2x Annealing Buffer RNAi

200 mM potassium acetate
4 mM magnesium acetate ▼
60 mM HEPES-KOH (pH 7.4) ▼

Electrotransfer Buffer RNAi

25 mM Tris ▼
192 mM glycine ▼
0.01% SDS▼
20% methanol▼

TBST (pH 7.4) RNAi

0.2% Tween-20
20 mM Tris-HCl
150 mM NaCl

RNAi in Mouse Oocytes and Early Embryos

These buffers are used in the preparation/maintenance of embryos, as well as introduction of dsDNA/RNA.

CZB Medium   RNAi

81.62 mM NaCl
4.83 mM KCl ▼
1.18 mM KH_2PO_4 ▼
1.18 mM $MgSO_4$ ▼
1.7 mM $CaCl_2$ ▼
25.12 mM $NaHCO_3$
31.3 mM sodium lactate

0.27 mM sodium pyruvate
1 mM L-glutamine
0.11 mM EDTA
3 mg/ml bovine serum albumin (BSA)
10 µg/ml gentamicin
10 µg/ml phenol red ▼

MEM/PVP Medium   RNAi

Bicarbonate-free minimal essential medium (Earle's salts) supplemented with sodium pyruvate (100 µg/ml), gentamicin (10 µg/ml), polyvinylpyrrolidone (PVP; 3 mg/ml) ▼ , and 25 mM HEPES (pH 7.2).

KSOM Medium   RNAi

95 mM NaCl
2.5 mM KCl ▼
0.35 mM KH_2PO_4 ▼
0.2 mM $MgSO_4$ ▼
10 mM sodium lactate
0.2 mM glucose
0.2 mM sodium pyruvate

25 mM $NaHCO_3$
1.71 mM $CaCl_2$ ▼
1 mM L-glutamine
0.01 mM EDTA
1 mg/ml BSA
10 µg/ml gentamicin

For RNA Isolation and Reverse Transcriptase–Polymerase Chain Reaction

Lysis Buffer   RNAi

4 M guanidine thiocyanate ▼
1 M 2-mercaptoethanol ▼
0.1 M Tris-HCl (pH 7.4)

RNA Resuspension Solution RNAi

40 mM Tris-HCl (pH 7.9)
10 mM NaCl
6 mM MgCl₂ ▼

For Oligo(dT)-primed Reverse Transcriptase RNAi

0.1 M dithiothreitol ▼
10 mM dNTPs
5x first-strand buffer
oligo(dT) (12–18) (500 ng/μl)
RNasin (40 units/μl)

For Random Hexamer-primed Reverse Transcription RNAi

0.1 M dithiothreitol ▼
10 mM dNTPs
5x first-strand buffer
random hexamers (1 μg/μl)
RNasin (40 units/μl)

For PCR Amplification RNAi

[α-³²P]dCTP (1 mCi/ml) (GE Healthcare) ▼
DNA polymerase (5 units/μl) (e.g., AmpliTaq, Applied Biosystems)
MgCl₂ (25 mM) ▼
10x PCR buffer
3′ and 5′ primers (2 μM each)

For Gel Electrophoresis RNAi

acrylamide (30%)/bis 37.5:1 solution ▼
ammonium persulfate (10%) ▼
6x electrophoresis sample-loading buffer
10x TBE ▼
TEMED ▼

NOTES

NOTES

SECTION 3
Imaging

Imaging

The following section contains procedures and recipes commonly used in live cell imaging such as the preparation and mounting of material, as well as the specific examination of subcellular compartments.

> *Caution:* See Cautions Appendix for appropriate handling of materials marked with ▼.

Preparation of Embryos for Live Cell Imaging

The following buffers are used in live imaging of mouse, chick, *Xenopus*, and *C. elegans* embryos. For specific protocols in preparing embryos for live imaging, refer to the laboratory manuals.

Early Embryos

Egg Salts Solution LCl

118 mM NaCl
40 mM KCl ▼
3.4 mM $MgCl_2$ ▼
3.4 mM $CaCl_2$ ▼
5 mM HEPES (pH 7.4)

Late-stage Embryos/Larvae

M9 Medium LCl

3 g of KH_2PO_4 ▼
6 g of Na_2HPO_4 ▼
5 g of NaCl
1 ml of 1 M $MgSO_4$ ▼
per liter (H_2O or phosphate-buffered saline)

Troubleshooting Guide for C. elegans *Live Imaging*   LCI

Problems	Solution
Animals move too much during imaging	Use less buffer or remove some buffer by capillary action with a piece of paper.
Larvae move and leave field of view	Place a thin coat of OP50 *E. coli* in the center of the coverslip. The worms will stay there and feed.
Embryos twitch or move too much	Use a temperature-controlled stage and a jacket surrounding the immersion objective to lower the temperature to 4–8°C.
Animals move	Use azide (3–10 mM) or levamisole (1 mM).
Focus drift during time-lapse microscopy	Keep the temperature in the room uniform using a good AC system (20–22°C).
Small focus drift during time-lapse microscopy	Adjust manually by moving the stage one or two clicks up or down until focus stabilizes (trial and error). Babysitting the embryos during imaging is sometimes necessary.
Out-of-focus fluorescence	Use confocal or deconvolution microscopy.
Animals do not stick to poly-L-lysine-coated coverslips	Use water instead of buffer. High salts may prevent binding. Alternatively, make fresh poly-L-lysine-coated coverslips.
Bubbles in the closed observation chamber	Press the coverslip gently with tweezers to displace the bubbles, or prepare a new chamber and gently place the coverslip using fine tweezers.

Examples of Live Cell Imaging of **Drosophila** *Tissues*   LCI

Tissue type	Subject	Approach
Egg chambers	border-cell migration	GFP fluorescence
	microtubules	*Tau*-GFP, tubulin-GFP, Exu-GFP
	mRNA transport	AlexaFluor546-RNA, FITC-RNA
Spermatogenesis	cell division	GFP, FRAP
Embryos	lipid droplet motility	DIC
	asymmetric cell division	*Tau*-GFP, *Src*-GFP
	asymmetric cell division	FRAP GFP-Aurora
	neuronal cell ablation	GFP
	marking clones	Caged-FITC-Dextran
	mRNA transport	AlexaFluor546-RNA
Larvae (dissected)	salivary gland chromosomes	DIC, GFP
	imaginal discs	GFP-Hh
	imaginal discs	Argos-GFP
Pupae	asymmetric cell division	GFP
Adults	eye in whole adult	Rhodapsin
	brain, calcium	Camgaroo

Optimal Conditions for Culturing Drosophila Tissues during Imaging LCI

Tissue	Optimized culture conditions	Imaging conditions
Early-stage to midstage egg chambers dissected from well-fed females	Halocarbon oil (Series 95) avoids dehydration/hypoxia	Early stages: Problem of activated and loose MT organization in aqueous media. Oil has higher refractive index than water. Good environment for injection.
Late-stage egg chambers (as above)	Grace's medium (Sigma-Aldrich) provides, ionic osmotic balance and nutrients	Late stages not susceptible to problems of early stages.
Embryos, dechorionated and dehydrated	Halocarbon oil (Series 700, RI similar to glycerol)	Prevents excess dehydration while avoiding hypoxia (which causes changes to the cell cycle). Good environment for injection.
	Halocarbon oil with Teflon breathable membrane	Better dehydration prevention for long-term development studies. Better bright-field imaging. Can help squash specimen for greater optical clarity (reduced spherical aberration).
	Aqueous medium	Water immersion objectives, which have longer working distance with high NA.

4% Paraformaldehyde   LCI

4% paraformaldehyde ▼ (10 g of paraformaldehyde) in 0.1 M NaPhosphate buffer (250 ml of 0.1 M NaPhosphate buffer)

Heat and stir at 60°C until clear. Adjust pH to 7.4.

NaPhosphate Buffer (0.1 M)   LCI

0.017 M NaPhosphate monobasic (3.0 g of NaPhosphate monobasic)
0.083 M NaPhosphate dibasic (14.2 g of NaPhosphate dibasic)

Dissolve in distilled H_2O (1250 ml of distilled H_2O).

Mammalian Embryo Culture

Dissection Medium   IND

The dissection medium is composed of 90% (v/v) DMEM/F-12 (GIBCO-BRL), 8% (v/v) heat-inactivated fetal bovine serum (GIBCO-BRL), 10 mM HEPES (Irvine Scientific), and 10 mM penicillin–streptomycin solution ▼ (Irvine Scientific). This medium must be heated to 37°C before use in a water bath.

Culture Medium   IND

The appropriate culture medium depends on the stage of the embryo. For younger embryos (embryonic day 5.5 [E5.5] to E7.5), the medium consists of 50% DMEM and 50% rat serum. For embryos between E7.5 and E9.5, DMEM/F12 is used instead of DMEM and the culture medium is supplemented with 10 mM HEPES and 10 mM penicillin-streptomycin solution (Irvine Scientific). Sterilize the medium by filtering through a 0.2-μm filter. At all stages, it is necessary to heat and equilibrate the pH of the medium by placing it in a tissue-culture incubator (5% CO_2, 37°C) for at least 1 hour.

The most important component of proper mammalian embryo culture is the use of high-quality rat serum. Although some commercial sources exist, homemade preparations consistently give superior results.

Composition of Physiological Solutions IND

Chemical components	Mouse ACSF		Chick ACSF		Zebrafish Danieau's 30x stock	
	g/l	mM	g/l	mM	g/l	mM
NaCl	6.95	119	7.29	125	101.7	1740
KCl ▼	0.18	2.5	0.37	5	1.56	21
MgCl$_2$•6H$_2$O ▼	0.26	1.3	0.41	2	–	–
CaCl$_2$•2H$_2$O ▼	0.37	2.5	0.30	2	–	–
NaH$_2$PO$_4$ ▼	0.12	1.0	–	–	–	–
KH$_2$PO$_4$ ▼	–	–	0.17	1.25	–	–
Glucose	1.98	11	3.6	20	–	–
HEPES (free acid)	4.76	20	5.48	23	35.75	150
MgSO$_4$•7H$_2$O ▼	–	–	–	–	2.96	12
Ca(NO$_3$)$_2$ ▼	–	–	–	–	4.25	18

Preparation of Chick Embryos

Howard Ringer's Solution (2 liters) IND

14.4 g of NaCl
0.45 g of CaCl$_2$•2H$_2$O ▼
0.74 g of KCl ▼

Mix in distilled water to 2 liters. Adjust pH to 7.4. Filter-sterilize and add to 125-ml sterile bottles.

Preparation of *Xenopus* Tadpoles for Imaging

MS-222 Anesthetic Solution IND

0.02% MS-222 (Sigma-Aldrich) in rearing solution. The solution can be stored for 1 month at 4°C.

Rearing solution
  58.2 mM NaCl
  0.67 mM KCl ▼
  0.34 mM CaNO$_3$ ▼
  0.78 mM MgSO$_4$ ▼
  50 mM HEPES

  Adjust to pH 7.4.

Embryo Development in *C. elegans*

Egg Buffer   IND

118 mM NaCl
48 mM KCl▼
3 mM CaCl$_2$ ▼
3 mm MgCl$_2$ ▼
5 mM HEPES (pH 7.2)

Imaging Fixed Tissue

The following buffers are used in fixing tissues/embryos for imaging. For specific fixation protocols, please refer to the reference materials.

MEMFA   IND

4% formaldehyde ▼
0.1 M MOPS ▼
2 mM EDTA
1 mM MgSO$_4$ (pH 7.6) ▼

Transfer 5–25 embryos to 5 ml of fixative. Fix overnight at 4°C or for 3 hours at room temperature. For long-term storage, dehydrate embryos and store them at –20°C in 100% ethanol ▼ or methanol ▼.

Dent's Fix   IND

80% methanol ▼
20% DMSO ▼

Transfer 5–25 embryos to 5 ml of fixative and quickly rinse at least twice over 5 minutes. Fix embryos overnight at –20°C and transfer to 100% methanol for long-term storage. F-actin can be visualized with phalloidin if acetone is substituted for methanol.

Gard's Fix   IND

This fixative is aldehyde-based ▼. Transfer 5–25 embryos to 5 ml of fixative. Fix overnight at 4°C. Process embryos immediately for best preservation of microtubule structures. Glutaraldehyde produces high fluorescence background that should be quenched with overnight incubation in NaBH$_4$ ▼ in PBS (phosphate-buffered saline).

TCA Fix IND

3% Trichloroacetic acid in PBS ▼

Transfer 5–25 embryos to 5 ml of fixative. Fix overnight at 4°C. Store embryos for up to 1 week in TCA or dehydrate in methanol for long-term storage. TCA does not fix lineage tracers such as fluorescently conjugated dextrans.

Preparation of Mice for Imaging

The following buffers are used for anesthesizing mice before live imaging. For specific anesthesia protocols, please refer to the laboratory manuals. Several commonly used anesthesias are listed below.

Ketamine/Xylazine Mixture LCI

12.5 mg/ml ketamine (2.6 ml [100 mg/ml bottle])
0.65 mg/ml xylazine (0.2 ml [100 mg/ml bottle])
in distilled H_2O (20 ml of H_2O)

Urethane Mixture LCI

20% urethane (1 g of urethane) ▼ in a 0.9% NaCl solution (5 ml of saline)

For Specific Examination of Neuronal Cells

Sterile Cortex Buffer LCI

125 mM NaCl (7.21 g of NaCl)
5 mM KCl (0.372 g of KCl) ▼
10 mM glucose (1.802 g of glucose)
10 mM HEPES (2.38 g of HEPES)
2 mM $CaCl_2$ (2 ml of 1 M $CaCl_2$) ▼
2 mM $MgSO_4$ (2 ml of 1 M $MgSO_4$) ▼
in distilled H_2O (1 liter of H_2O)
Adjust pH to 7.4.

Intracellular Imaging

The following buffers are used for specific live visualization of intracellular components, nucleic acid, filaments, and other organelles.

Visualizing Nucleic Acid

Chromosomes can be visualized in living cells simply by adding Hoechst buffer to cell culture medium.

Stock Solution of Hoechst 33342   LCI

Frozen Stock
10 mg/ml Hoechst 33342 in double-distilled H_2O. Store at $-30°C$.

Working Solution
100 μg/ml Hoechst 33342 in double-distilled H_2O. Store in the dark at 4°C.
This Hoechst 33342 solution can be stored for 6 months at 4°C.

Visualizing Protein

Proteins are commonly visualized by fluorescent tagging. To visualize fluorescently labeled protein live, observation culture medium can be used to replace cell culture medium.

Observation Culture Medium   LCI

DME medium without phenol red
10% fetal bovine serum (FBS)
20 mM HEPES buffer (pH 7.3)
80 μg/ml kanamycin sulfate ▼

DME medium can be replaced with any medium, but when observing red fluorescence (e.g., from rhodamine), phenol red can cause background and should be removed or reduced.

Visualizing the Cytoskeleton

GFP-tagged actin can be used to visualize the cytoskeleton. Transfer of DNA can be conducted using the calcium phosphate method. In this case, the following buffers can be used and cells can be visualized by HEPES-buffered Tyrode's solution.

2× **BBS (BES-buffered saline)** LCI

280 mM NaCl
1.5 mM Na_2HPO_4 ▼
50 mM BES buffer (pH 7.1) (Sigma-Aldrich)

It is convenient to make up 100 ml of this solution. Dissolve the components in 80 ml of distilled H_2O, adjust the pH to 7.1 with NaOH, and filter-sterilize.

HBS (HEPES-buffered Saline) LCI

135 mM NaCl
1 mM Na_2HPO_4 ▼
4 mM KCl ▼
2 mM $CaCl_2$ ▼
20 mM HEPES buffer (Sigma-Aldrich)

Make up 100 ml, adjust the pH to 7.5, and filter-sterilize.

HEPES-buffered Tyrode's Solution LCI

119 mM NaCl
5 mM KCl ▼
25 mM HEPES buffer
2 mM $CaCl_2$ ▼
2 mM $MgCl_2$ ▼
6 g/liter glucose

Adjust pH to 7.4 with NaOH.

For Imaging Actin from Slices of Transgenic Mouse Brain

Artificial Cerebrospinal Fluid (ACSF) LCI

124 mM NaCl
2.5 mM KCl ▼
2.0 mM $MgSO_4$ ▼
1.25 mM KH_2PO_4 ▼
26 mM $NaHCO_3$
10 mM glucose
4 mM sucrose
2.5 mM $CaCl_2$ ▼

When the components are first mixed, the ACSF solution may be slightly milky, but it should clear when aerated with 95% O_2, 5% CO_2, after which the pH of the medium should be 7.4.

Imaging Intermediate Filament Proteins in Living Cells

Homogenization Buffer LCI

50 mM Tris-HCl (pH 7.4)
5 mM $MgCl_2$ ▼
0.1% 2-mercaptoethanol ▼
1 mM phenylmethylsulfonyl fluoride (PMSF) ▼

Extraction Buffer LCI

8 M urea ▼
50 mM Tris-HCl (pH 7.4)
5 mM EGTA
0.2% 2-mercaptoethanol ▼
1 mM PMSF ▼

Column Buffer LCI

6 M urea ▼
8 mM sodium phosphate (pH 7.2) ▼
0.14 M NaCl
1 mM dithiothreitol (DTT) ▼
0.1 mM PMSF ▼

Assembly Buffer LCI

6 mM sodium/potassium phosphate (pH 7.4) ▼
0.17 M NaCl
3 mM KCl ▼
0.2% 2-mercaptoethanol ▼
0.2 mM PMSF ▼

Purification of Intermediate Filament Proteins from Bacterial Inclusion Bodies

Intermediate filament (IF) proteins can be expressed in recombinant bacterial systems, purified, and labeled separately. The following buffers may be used to purify IF proteins from bacterial inclusion bodies.

Homogenization Buffer LCI

50 mM Tris-HCl (pH 8)
25% sucrose
1 mM EDTA

Detergent Buffer LCl

25 mM Tris-HCl (pH 8)
0.2 M NaCl
1% Nonidet P-40
1% deoxycholate ▼
1 mM EDTA

Wash Buffer LCl

TNE buffer
0.5% Nonidet P-40
3 mM MgCl$_2$ ▼

Column Buffer LCl

10 mM Tris-HCl (pH 8)
8 M urea ▼
2 mM EDTA
10 mM DTT ▼

TNE LCl

10 mM Tris-HCl (pH 8)
100 mM NaCl
1 mM EDTA

Disassembly Buffer LCl

8 M urea ▼
5 mM sodium phosphate (pH 7.2) ▼
0.2% 2-mercaptoethanol ▼
2 mM PMSF ▼

Dialysis Buffer LCl

5 mM sodium phosphate (pH 7.4) ▼
0.2% 2-mercaptoethanol ▼
0.2 mM PMSF ▼

Sephadex Column Buffer LCl

5 mM sodium phosphate (pH 7.4) ▼
0.2% 2-mercaptoethanol ▼
0.2 mM PMSF ▼

Imaging Medium LCI

Phenol-red-free cell growth medium (e.g., RPMI or DMEM)
10% fetal bovine serum
2 mM glutamine
25 mM HEPES (pH 7.4)

Imaging Living Neurons

Preparing Slice Cultures IND

Add 1 ml of Organotypic slice culture medium to a 35-mm petri dish and place a Millicell-CM membrane insert on the medium.

Organotypic Slice Culture Medium

Minimal essential medium (MEM) with Hanks' salts, no glutamine (GIBCO)

25% horse serum (Hyclone)

30 mM glucose

5 mM NaHCO$_3$

30 mM HEPES

2.5 mM MgSO$_4$ ▼

2 mM CaCl$_2$ ▼

3 mM L-glutamine

1 mg/liter insulin (Sigma-Aldrich)

10 ml/liter penicillin-streptomycin (Sigma-Aldrich P0781)

Adjust pH to 7.2 at 35°C.

Note: The membrane becomes clear when wet, facilitating imaging from below. The plastic ring on standard inserts gets in the way of petri dish lids, electrodes, and large objective lenses (upright scope), so this was initially cut off with a hot nichrome wire. Millipore produces "low height" inserts (PICM0RG50) to avoid this problem.

Ballistic Delivery of Dyes

Particle-mediated ballistic delivery of fluorescent dyes has recently been used to label neuronal populations. Several buffers used in this process are listed below.

Lipophilic Dyes IND

For single-color labeling, dissolve 3 mg of lipophilic dye ▼ such as DiI, DiO, or DiD in 100 μl of methylene chloride ▼ (Sigma-Aldrich). For multicolor labeling, dissolve 7 mg of DiI, DiO, and DiD in 70 μl of methylene chloride in three separate microfuge tubes. Mix various proportions of the different dyes to obtain, for example, seven different dye combinations in seven tubes as follows: (a) 30 μl of DiI, (b) 30 μl of DiO, (c) 30 μl of DiD, (d) 15 μl of DiI: 15 μl of DiO, (e) 15 μl of DiI: 15 μl of DiD, (f) 15 μl of DiO: 15 μl of DiD, and (g) 10 μl of DiI: 10 μl of DiO: 10 μl of DiD. Last, add 70 μl of methylene chloride to each tube. The final concentration for each dye combination is 3 mg in 100 μl of methylene chloride.

Slice Solution IND

126 mM NaCl
2.5 mM KCl ▼
1.25 mM NaH_2PO_4 ▼
2 mM $MgCl_2$ ▼
2 mM $CaCl_2$ ▼
10 mM glucose
26 mM NaH_2CO_3 (pH 7.4)

Gas the solution with 5% CO_2 and 95% O_2 at room temperature (23–24°C).

Note: Rats are anesthetized with pentobarbital sodium ▼ (55 mg/kg) and decapitated. Coronal brain slices (300 μm) are cut with a vibratome in a cutting solution containing 2.5 mM KCl, 1.25 mM NaH_2PO_4, 10 mM $MgSO_4$ ▼, 0.5 mM $CaCl_2$, 10 mM glucose, 26 mM $NaHCO_3$, and 230 mM sucrose. Slices containing the hippocampus are incubated in the slice solution gassed with 5% CO_2 and 95% O_2, for 1 hour, then transferred to the recording chamber.

5% Biocytin Solution   IND

Using two patch pipettes, load 0.5% biocytin into each cell type under study. First, patch an interneuron with a biocytin-containing pipette and then patch an astrocyte with another biocytin-containing pipette. (0.5% Biocytin: Dissolve biocytin hydrochloride [Sigma-Aldrich] in the pipette solution. Use freshly prepared solution.)

Pipette solution

    120 mM K-gluconate ▼

    10 mM KCl ▼

    1 mM $MgCl_2$ ▼

    10 mM HEPES

    0.1 mM EGTA

    0.025 mM $CaCl_2$ ▼

    1 mM ATP

    0.2 mM GTP

    4 mM glucose

    Adjust pH to 7.2 with KOH ▼.

Fluorescent Dye or Plasmid DNA Labeling Solution   IND

DiI is 0.01–0.05% in absolute ethanol; fluorescent dextran is 5% (w/v) in 0.1 M phosphate buffer or water; plasmid DNA is 0.1–5.0 µg/µl in water or elution buffer. An endotoxin-free plasmid purification kit is recommended for plasmid preparation. Dye solutions can all be stored for 1 month at –20°C, although fluorescent dyes are generally best made fresh from powder. Care should be taken to avoid exposure of fluorescent dyes to light at all times.

Imaging Synaptic Vesicles

Normal Frog Ringer's Solution   IND

115 mM NaCl

2 mM KCl ▼

1.8 mM $CaCl_2$ ▼

5 mM HEPES (pH 7.2, osmolarity ~225 mOsm)

High [K^+] solution used for FM labeling contains 60 mM KCl and 57 mM NaCl. Dye concentrations are usually 2–10 µM for FM 1-43 and for FM 4-64, and 25–40 µM for FM 2-10.

Imaging Exocytosis

Depolarizing Buffer   IND

95 mM NaCl
25 mM KCl ▼
2.5 mM CaCl$_2$ ▼
1 mM MgCl$_2$ ▼
10 mM HEPES (pH 7.2)

Standard Buffer   IND

120 mM NaCl
2.5 mM KCl ▼
2.5 mM CaCl$_2$ ▼
1 mM MgCl$_2$ ▼
10 mM HEPES (pH 7.2)

Culture Medium   IND

45% DMEM ▼
45% Ham's F-12
10% fetal calf serum
1 unit/ml penicillin-streptomycin ▼

All are from GIBCO-BRL.

Low Calcium Buffer   IND

120 mM NaCl
2.5 mM KCl ▼
0.5 mM CaCl$_2$ ▼
1.25 mM EGTA
1 mM MgCl$_2$ ▼
10 mM HEPES (pH 7.2)

General Information for Imaging

Listed below are tables with general information.

Some Mammalian Cells Used for Microinjection LCI

Name	Cell type	Origin	ATTC number
CHO-K1	epithelial	ovary; *Cricetulus griseus* (Chinese hamster)	CCL-61
CV-1	fibroblastic	kidney; normal; *Cercopithecus aethiops* (African green monkey)	CCL-70
COS-7	fibroblastic	SV40 transformed derivative of CV-1	CRL-1651
3T3-Swiss	fibroblastic	embryo; *Mus musculus* (mouse)	CCL-92
NRK	epithelial	kidney; normal; *Rattus norvegicus* (rat)	CRL-6509
PtK1	epithelial	kidney; normal; *Potorous tridactylis* (potoroo)	CRL-6493
LLCPK	epithelial	kidney; normal; porcine	

Objective Lens Magnification Numerical Aperature, Pixel Size, and Signal Brightness LCI

Magnification	NA[a]	d_{xy} (µm)	P_{xy} (µm)	Brightness
10	0.3	1.037	5.2	0.11
20	0.5	0.622	6.2	0.20
40	0.7	0.444	8.9	0.20
63	1.32	0.236	7.4	1.00
100	1.4	0.222	11.1	0.50

[a]The objective lenses are from Leica.

Application of Photobleaching Methods LCI

Application	Method
Descriptive[a]	
Detection of mobile and immobile fractions	FRAP, iFRAP, FLIP, photoactivation, or spatial analysis
Comparison of kinetics	FRAP, iFRAP, FLIP, or photoactivation
Compartment connectivity	FLIP
	spatial analysis
	photoactivation
Molecular dynamics analysis	
Diffusion coefficient	FRAP
	photoactivation
	spatial analysis
Association and dissociation kinetics	FRAP
	iFRAP
	FLIP
	photoactivation

[a]Because this information is not linked to a model, and therefore to biophysical or biochemical properties, it is referred to as "descriptive," in contrast to molecular dynamics analysis.

Example of FRET Fluorophore Pairs LCI

Donor	Excitation$_{Donor}$ (nm)	Emission$_{Donor}$ (nm)	Acceptor	Excitation$_{Acceptor}$ (nm)	Emission$_{Acceptor}$ (nm)
CFP	440	480	YFP	520	535
BFP	365	460	GFP	488	535
CFP	440	480	dsRed1	560	610
FITC	488	535	Cy3	525	595
Cy3	525	595	Cy5	633	695
GFP	488	535	Rhodamine	543	595

NOTES

Protein Analysis/Proteomics

Protein Analysis/Proteomics

The following section covers buffers used in protein analysis and investigation of protein–protein interactions. This section is presented in two parts: (1) basic methods in protein analysis for proteomics and (2) advanced protocols in identifying protein–protein interactions for proteomics. Areas covered include peptide sequencing, analysis of peptide/DNA complexes, and two-hybrid and library screening.

> *Caution:* See Cautions Appendix for appropriate handling of materials marked with ▼.

Basic Methods in Protein Analysis for Proteomics

Chromatography

The following sections include solutions used in various chromatographic purification protocols.

■ *Size Exclusion Chromatography*

Denaturing Buffer Containing Urea   PUR

8 M urea ▼
5 mM dithiothreitol ▼
150 mM NaCl
50 mM Tris-Cl (pH 7.5)

Warm the solution to 40°C to dissolve the urea. Adjust the pH to 7.5 if necessary.

Denaturing Buffer Containing Guanidine Hydrochloride PUR

6 M guanidine hydrochloride ▼
5 mM dithiothreitol ▼
50 mM Tris-Cl (pH 7.5)

Warm the solution to 40°C to dissolve the guanidine hydrochloride. Adjust the pH to 7.5 if necessary.

Peptide Elution Buffers PUR

20% acetonitrile ▼ /0.1% TFA ▼ in H_2O
0.1% TFA in H_2O
or
0.1% acetic acid ▼ in H_2O

▪ Hydrophobic Interaction Chromatography

High-density Phenyl-substituted Matrix Column PUR

Suitable columns include the POROS HP2 (M) chromatography column (Applied Biosystems), SOURCE 15 PHE PE (100 × 4.6 mm), or Bio-Gel TSK Phenyl-5PW HIC column (75 × 7.5 mm I.D. 1000 Å) (Bio-Rad).

Dimethyl Pimelimidate (DMP) Solution PUR

Prepare 10 ml of 20 mM DMP (Pierce) in 0.1 M triethanolamine ▼ (pH 9.5; the pH of the solution decreases to ~8.0 due to the acidity of DMP). This reagent must be prepared fresh immediately before use.

Column Regeneration Buffer PUR

10 mM Tris-HCl (pH 7.6)
1 mM EDTA
2.5 M NaCl
1% Nonidet P-40 (NP-40)

NP-40 will never completely dissolve. Swirl just before use.

Column Storage Buffer PUR

10 mM Tris-HCl (pH 7.6)
1 mM EDTA
0.3 M NaCl
0.04% sodium azide ▼

Add sodium azide fresh from a 4% stock solution just before use.

HEMGN Buffer PUR

25 mM HEPES (K⁺, pH 7.6)
0.1 mM EDTA
2.5 mM $MgCl_2$ ▼
10% glycerol
0.025% Nonidet P-40
1 mM dithiothreitol (DTT) ▼

Add DTT fresh immediately before use. Store the buffer at 4°C.

■ Immobilized Metal-ion Affinity Chromatography

Supplemented Buffer P PUR

50 mM sodium phosphate (pH 7.4) ▼
0.5 M NaCl
0.5 mg/ml lysozyme ▼
1 mM PMSF ▼
1.7 units/ml Benzonase (Merck)
1 mM $MgCl_2$ ▼

Add the lysozyme, PMSF, and Benzonase just before use.

Binding Buffer PUR

20 mM sodium phosphate (pH 7.4) ▼
0.5 M NaCl
20 mM imidazole ▼

Elution Buffer PUR

20 mM sodium phosphate (pH 7.4) ▼
0.5 M NaCl
500 mM imidazole ▼

Inclusion Body Wash Buffer   PUR

2 M urea ▼
500 mM NaCl
2% Triton X-100 ▼
20 mM Tris-HCl (pH 8.0)

Refolding Buffer   PUR

20 mM imidazole ▼
500 mM NaCl
1 mM 2-mercaptoethanol ▼
20 mM Tris-HCl (pH 8.0)

Urea Binding Buffer   PUR

For preparation of buffers containing urea, weigh the denaturant, add 2 M imidazole ▼, 1 M Tris-HCl (pH 8.0), and H_2O to 90% of the final volume. Heat the container, which is cooled by the solubilization process of the denaturant, to room temperature in a water bath at 45°C while stirring. Avoid high temperatures. The denaturant will dissolve within minutes. Alternatively, stir overnight at room temperature. Adjust the pH, add H_2O to the final volume, and filter the solution. The pH adjustment must be done at room temperature. Add 2-mercaptoethanol ▼ immediately before use.

■ Chromatofocusing

The following section includes a solution used in purifying proteins by chromatofocusing.

Acid Solutions   PUR

2 M acetic acid ▼
2 M HCl ▼
iminodiacetic acid (saturated solution) ▼

These solutions are used for adjusting the pH of start and elution buffers.

▪ *Two-dimensional Gel Electrophoresis*

The following section includes solutions for preparation and visualization of proteins by two-dimensional (2D) electrophoresis.

Thiourea/Urea Lysis Buffer   PUR

2 M thiourea ▼
7 M urea ▼
4% (w/v) CHAPS ▼
1% (w/v) DTT ▼
2% (v/v) carrier ampholytes (pH 3–10)
10 mM Pefabloc proteinase inhibitor

To prepare 50 ml of thiourea/urea lysis buffer, dissolve 22.0 g of urea (GE Healthcare) in deionized H_2O, add 8.0 g of thiourea (Sigma-Aldrich), and adjust the volume to 50 ml with deionized H_2O. Add 0.5 g of Serdolit MB-1 mixed-bed ion-exchange resin (Serva), stir for 10 minutes, and filter. To 48 ml of the urea solution add 2.0 g of CHAPS, 1.0 ml of Pharmalyte (pH 3–10; GE Healthcare), 0.5 g of DTT, and, immediately before use, 50 mg of Pefabloc proteinase inhibitor (VWR).

Urea Lysis Buffer   PUR

9.5 M urea ▼
1% (w/v) DTT ▼
2% (w/v) CHAPS ▼
2% (v/v) carrier ampholytes (pH 3–10)
10 mM Pefabloc proteinase inhibitor

To prepare 50 ml of urea lysis buffer, dissolve 30.0 g of urea (GE Healthcare) in deionized H_2O and adjust the volume to 50 ml. Add 0.5 g of Serdolit MB-1, stir for 10 minutes, and filter. Add 1.0 g of CHAPS, 0.5 g of DTT, and 1.0 ml of Pharmalyte (pH 3–10) to 48 ml of the filtered urea solution. If necessary, add 50 mg of Pefabloc proteinase inhibitor immediately before use.

IMPORTANT: Lysis buffer must be freshly prepared. Alternatively, make 1-ml aliquots and store at –80°C for up to several months. Do not refreeze lysis buffer that has already been thawed once! Never heat urea solutions above 37°C. Otherwise, protein carbamylation may occur.

Reswelling Buffer PUR

8 M urea ▼
0.4% (w/v) dithiothreitol (DTT) ▼
1% (w/v) CHAPS ▼
2.5% Pharmalyte (pH 3–10)

To prepare 50 ml of reswelling buffer, dissolve 25.0 g of urea in deionized H_2O and adjust the volume to 50 ml. Add 0.5 g of Serdolit MB-1 mixed-bed ion-exchange resin (Serva), stir for 10 minutes, and filter. Add 0.5 g of CHAPS, 0.2 g of DTT, and 1.25 ml of Pharmalyte (pH range 3–10) (GE Healthcare) to 48 ml of the urea solution.

IPG DryStrip Rehydration Buffer PUR

8 M urea ▼ (or 6 M urea + 2 M thiourea ▼ ; see Tip below)
0.4% (w/v) dithiothreitol (DTT) ▼
1% (w/v) CHAPS ▼
2.5% Pharmalyte (pH 3–10)

To prepare 50 ml of reswelling buffer, dissolve 25.0 g of urea in deionized H_2O and adjust the volume to 50 ml. Add 0.5 g of Serdolit MB-1 mixed-bed ion-exchange resin (Serva), stir for 10 minutes, and filter. Add 0.5 g of CHAPS, 0.2 g of DTT, and 1.25 ml of Pharmalyte (pH 3–10) to 48 ml of the urea solution.

Experimental Tip: For isoelectric focusing (IEF) of hydrophobic proteins, 6 M urea with 2 M thiourea is superior to 8 M urea. For IEF of basic proteins, the use of DeStreak instead of IPG DryStrip rehydration buffer is preferable.

Equilibration Buffer PUR

6 M urea ▼
30% (w/v) glycerol
2% (w/v) SDS ▼ in 0.05 M Tris-HCl buffer (pH 8.8)

To make 500 ml, combine 180 g of urea, 150 g of glycerol, 10 g of SDS, and 16.7 ml of Tris-HCl buffer. Dissolve in deionized H_2O and adjust the volume to 500 ml. This buffer can be stored for up to 2 weeks at room temperature.

Towbin Transfer Buffer PROT

Tris-base	15.1 g
glycine ▼	72.1 g
methanol ▼	500 ml
SDS (optional) ▼	5.0 g
H_2O	to 5 liters

10% methanol is recommended for binding proteins to nitrocellulose membranes. Up to 20% methanol can be included. Methanol helps remove SDS from the proteins and improves the binding of the proteins to the membrane. SDS is optional, but it can be included to help improve the transfer of large proteins.

CAPs Transfer Buffer PROT

Add 2.2 g of 3-(cyclohexyl-amino)-1-propanesulfonic acid ▼ to approximately 600 ml of H_2O. Adjust the pH to 11 with concentrated NaOH ▼ and adjust the volume to 1 liter with H_2O.

Anode Stabilization Media PUR

glycerol (14.5% w/w)
hydroxypropylmethyl cellulose (HPMC) (0.12% w/w)
100 mM H_2SO_4 ▼
8 M urea ▼

For 68.9 g = approximately 60 ml, weigh out 6 g of H_2SO_4 (1 M). Add 9 g of H_2O and 25 g of HPMC/glycerol/H_2O stock (40%/40%/20% w/w/w). Add 28.9 g of urea. Dissolve by stirring in lukewarm water. Actual consumption is approximately 10 g/hr.

Cathode Stabilization Media PUR

glycerol (14.5% w/w)
HPMC (0.12% w/w)
100 mM NaOH ▼
8 M urea ▼

For 137.8 g = approximately 120 ml, weigh out 12 g of 1 M NaOH. Add 18 g of H_2O and 50 g of HPMC/glycerol/H_2O stock (40%/40%/20% w/w/w). Add 57.8 g of urea. Dissolve by stirring in lukewarm water. Actual consumption is approximately 20 g/hr.

Counterflow Media   PUR

glycerol (14.5% w/w)
HPMC (0.12% w/w)
8 M urea ▼

For 330.5 g = approximately 288 ml, weigh out 120 g of HPMC/glycerol/H_2O stock (40%/40%/20% w/w/w). Add 72 g of H_2O and 138.5 g of urea. Dissolve by stirring in lukewarm water. Actual consumption is approximately 46 g/hr.

Electrolyte Cathode Circuit   PUR

NaOH (100 mM) ▼

For 400 g, weigh out 40 g of NaOH (1 N). Add 360 g of H_2O. Dissolve by stirring. Four hundred grams is the minimum amount needed to run workday experiments. Use larger amounts for longer-term experiments, i.e., >12 hours. Actual consumption is none.

Separation Media 1   PUR

glycerol (14.5% w/w)
HPMC (0.12% w/w)
ProLyte 1 (11.6% w/w)
8 M urea ▼

For 68.9 g = approximately 60 ml, weigh out 8 g of ProLyte 1. Add 7 g of H_2O and 25 g of HPMC/glycerol/H_2O stock (40%/40%/20% w/w/w). Add 28.9 g of urea. Dissolve by stirring in lukewarm water. pH = 4.83 ± 0.10; conductivity 207 ± 15 μS/cm. Actual consumption is approximately 10 g/hr.

Separation Media 2   PUR

glycerol (14.5% w/w)
HPMC (0.12% w/w)
ProLyte 2 (19.4% w/w)
8 M urea ▼

For 137.8 g = approximately 120 ml, weigh out 26.7 g of ProLyte 2. Add 3.3 g of H_2O and 50 g of HPMC/glycerol/H_2O stock (40%/40%/20% w/w/w). Add 57.8 g of urea. Dissolve by stirring in lukewarm water. pH = 7.54 ± 0.10; conductivity 347 ± 20 μS/cm. Actual consumption is approximately 20 g/hr.

Separation Media 3 PUR

glycerol (14.5% w/w)
HPMC (0.12% w/w)
ProLyte 3 (14.5% w/w)
8 M urea ▼

Production of 68.9 g = approximately 60 m, weigh in 10 g of ProLyte 3. Add 5 g of H_2O and 25 g of HPMC/glycerol/H_2O stock (40%/40%/20% w/w/w). Add 28.9 g of urea. Dissolve by stirring in lukewarm water. pH = 10.26 ± 0.17; conductivity 235 ± 20 μS/cm. Actual consumption is approximately 10 g/hr.

▪ Carbohydrate Analysis

The following solution is used in protocols for analyzing carbohydrates from glycoproteins.

Sugar Mix PUR

Prepare a 20-mg/ml stock solution in H_2O of each sugar separately. Combine 0.1 ml of each required solution. Use 50 μl of this sugar mix whenever an analysis is performed to determine response factors. Store all solutions at −20°C.

Standards for sugar mix (vary depending on the source of glycans)

meso-erythritol	2-deoxy-D-ribose
L-rhamnose	L-fucose
D-ribose	L-arabinose
D-xylose	2-deoxy-D-glucose
D-allose	D-mannose
D-galactose	D-glucose
myo-inositol	N-acetyl-D-glucosamine
N-acetyl-D-galactosamine	

Extraction of Proteins

The following solutions are used in the extraction of proteins from plant, mammalian, and bacterial cells.

▪ *From Mammalian Cells*

Homogenization Buffer A PROT

50 mM Tris-Cl (pH 7.5)
2 mM EDTA
150 mM NaCl
0.5 mM DTT ▼

Homogenization Buffer B PROT

50 mM Tris-Cl (pH 7.5)
10% (v/v) glycerol (or 0.25 M sucrose)
5 mM magnesium acetate
0.2 mM EDTA
0.5 mM DTT ▼
1.0 mM phenylmethylsulfonyl fluoride (PMSF) ▼

▪ *From Cultured Cells for IP*

NP-40 Lysis Buffer PROT

150 mM NaCl
1.0% Nonidet P-40
50 mM Tris-Cl (pH 7.4)

Probably the most widely used lysis buffer, it relies on the nonionic detergent NP-40 as the major solubilizing agent, which can be replaced by Triton X-100 with similar results. Variations include lowering the detergent concentration or using alternate detergents such as digitonin, saponin, or CHAPS.

RIPA Lysis Buffer PROT

150 mM NaCl
1% Nonidet P-40
0.1% SDS ▼
0.5% sodium deoxycholate ▼
50 mM Tris-Cl (pH 7.4)

A much harsher denaturing lysis buffer than NP-40, due to the inclusion of two ionic detergents (SDS and sodium deoxycholate). In addition to releasing most proteins from cultured cells, RIPA lysis buffer disrupts most weak non-covalent protein–protein interactions.

▪ From Cultured Animal Cells, Yeast, and Bacteria for Immunoblotting

1x Laemmli Sample Buffer PROT

2% SDS ▼
10% glycerol
60 mM Tris-Cl (pH 3.8)
0.01% bromophenol blue ▼

It is often convenient to prepare Laemmli sample buffer as a 2x or 5x stock. Just before use, add DTT to a final concentration of 100 mM.

▪ Preparation of Cellular and Subcellular Extracts

Cytoskeleton Solubilization Buffer PROT

5% SDS ▼
10 mM sodium phosphate (pH 7.4) ▼

For nonreducing buffer, dissolve 0.5 g of SDS in 5 ml of 20 mM sodium phosphate buffer, pH 7.4. Add H_2O to 10 ml (final volume). For denaturing buffer, add 1 ml of β-mercaptoethanol ▼ . Adjust H_2O appropriately.

1x O'Farrell Lysis Buffer PROT

Dissolve 5.7 g of ultrapure electrophoretic-grade urea ▼ , 0.2 ml of Nonidet P-40, 0.2 ml of ampholines (0.16 ml at pH 5–7 and 0.04 ml at pH 3–10), and 0.5 ml of β-mercaptoethanol ▼ in H_2O. Adjust final volume to 10 ml with H_2O. Solution may be warmed to facilitate solubilization. Divide into 1-ml aliquots and store at –70°C.

4x PIPES Buffer (piperazine-N,N´-bis[2-ethanesulfonic acid]) PROT

Dissolve 103 g of sucrose and 5.8 g of NaCl in 150 ml of H_2O. Dissolve 3 g of PIPES in a small volume of 1 M NaOH ▼ . Mix the PIPES solution with the sucrose/NaCl solution. Add 0.64 g of $MgCl_2 \cdot 6H_2O$ ▼ to the mixture. Adjust final volume to 250 ml and sterilize by passing through a 0.45-μm sterile filter. Store in the dark at 4°C or aliquot and freeze (stable for 2 months at 4°C). Use an aseptic technique when diluting the 4x PIPES buffer for preparation of 1x working solutions or store as single-use aliquots to avoid contamination.

■ *For 2D Gel Electrophoresis*

Extraction Solution for Rat Liver Protein Extract PROT

7 M urea▼	4.2 g
2 M thiourea▼	1.52 g
2% (w/v) ASB-14 (Calbiochem)	0.2 g
0.5% (v/v) Pharmalyte (pH 3–10) (AP Biotech)	50 µl
H₂O	to 10 ml

Prepare the solution fresh or store at –80°C. Cool on ice before use. Can be stored for up to 6 months at –80°C.

Extraction Buffer for Eukaryotic Lysates PROT

9 M urea▼	5.4 g
4% CHAPS▼	0.4 g
0.5% IPG buffer (pH 3–10) (AP Biotech)	50.0 µl
50 mM DTT ▼	0.077 g
H₂O	to 10 ml

Extraction Solution for E. coli Lysates PROT

8 M urea ▼	4.8 g
4% (w/v) CHAPS▼	400 mg
20 mM Tris-base	100 µl of 2 M Tris ▼
1 mM MgCl₂ ▼	100 µl of 100 mM MgCl₂ · 6H₂O ▼
H₂O	to 10 ml

Prepare the solution fresh or store at –80°C. Cool on ice before use.

■ *In Situ Peptide Mapping*

The following solutions are used for performing peptide maps, using the protocol that relies on initial separation by gel electrophoresis, in which peptide fragmentation is performed in situ and the partial digest is separated in a second gel.

Proteinase Digestion Buffer PROT

0.125 M Tris-CI (pH 6.8)
0.1% SDS ▼
20% glycerol
0.005% bromophenol blue ▼

Proteinase Solution PROT

0.1% SDS ▼
10% glycerol
0.005% bromophenol blue ▼
1–100 μg/ml proteinase in 0.125 M Tris-HCl (pH 6.8)

Endoproteinase Glu-C (or V8 protease), which can tolerate high concentrations of SDS, is recommended for the Cleveland method.

■ **By Hydroxylamine Fragmentation**

Cleavage Buffer PROT

2 M hydroxylamine HCl ▼
4.3 M guanidine-HCl ▼
0.2 M potassium carbonate (pH 9.0) ▼

Dissolve 3.5 g of hydroxylamine (Analar grade, BDH) in 12 ml of prechilled 6 M aqueous guanidine-HCl (use an ice bath), and then add 2 ml of 50% (w/v) NaOH ▼ slowly with vigorous stirring (magnetic stir bar) followed by 5 ml of 1 M potassium carbonate. Adjust the pH to 9.0 with 50% NaOH and then adjust the volume to 25 ml with 6 M guanidine-HCl.

The inclusion of guanidine-HCl is reported to improve the efficiency of Asn-Gly cleavage.

Advanced Protocols in Identifying Protein–Protein Interactions for Proteomics

Purification of Proteins and Inhibitors from Cell Cultures

The following buffers are used in purifying inhibitors from proteins by ion-exchange chromatography and gel filtration.

Ammonium Sulfate Cut Dilution Buffer (ASCDB) PPI2

20 mM Tris-HCl (pH 7.2)
1 mM DTT ▼

Prepare fresh, maintain at 4°C.

Dialysis Buffer PPI2

ASCDB supplemented with 20% glycerol
Prepare fresh, maintain at 4°C.

Buffer A (FPLC Mono Q)   PPI2

50 mM Tris-HCl (pH 7.2)
1 mM DTT ▼
Prepare fresh, maintain at 4°C.

Buffer B (FPLC Mono Q)   PPI2

50 mM Tris-HCl (pH 7.2)
1 M NaCl
1 mM DTT ▼
Prepare fresh, maintain at 4°C.

Disruption Buffer   PPI2

40 mM HEPES (pH 7.2)
10 mM EDTA
2 mM DTT ▼
5 µg/ml pepstatin ▼
10 µg/ml leupeptin ▼
5 µg/ml chymostatin ▼
Prepare fresh, maintain at 4°C.

Ubiquitination Buffer   PPI2

40 mM Tris-HCl (pH 7.6)
1 mg/ml rcm-BSA (bovine serum albumin, reduced, carboxymethylated)
1 mM DTT ▼
5 mM $MgCl_2$ ▼
10 mM phosphocreatine
50 mg/ml creatine phosphokinase,
50 µM ubiquitin
1 µM ubiquitin aldehyde (Calbiochem or Sigma-Aldrich)
1 pmole of E1
5 pmole of E2-C,
1–2 pmole (~105 cpm) of ^{125}I-labeled cyclin B (13–91)/protein A
Prepare fresh, maintain at 4°C.

▪ Coimmunoprecipitation

The following buffers are provided and used in the identification of associated protein complexes. In the example given in reference material, the following buffers are used to identify novel proteins.

EBC Lysis Buffer   PPI2

50 mM Tris (pH 8.0) ▼
120 mM NaCl
0.5% Nonidet P-40
5 µg/ml leupeptin ▼ *
10 µg/ml aprotinin ▼ *
50 µg/ml phenylmethylsulfonyl fluoride (PMSF) ▼ *
0.2 mM sodium orthovanadate ▼ *
100 mM NaF ▼ *

*Add just before use.

Laemmli Sample Buffer   PPI2

2% (w/v) SDS ▼
10% (v/v) glycerol
100 mM DTT ▼
60 mM Tris-HCl (pH 6.8)
0.001% (w/v) bromophenol blue ▼

NETN   PPI2

420 mM Tris (pH 8.0) ▼
41 mM EDTA
40.5% Nonidet P-40
4100 mM NaCl

▪ Using FLAG Epitope-tagged Proteins

IEF Buffer   PROT

9 M urea ▼
324 mM dithiothreitol (DTT) ▼
2% Pharmalyte (pH 3–10) (GE Healthcare)
0.5% Triton X-100 ▼

Lysis Buffer PROT

1% Triton X-100 ▼
10% glycerol
150 mM NaCl
20 mM Tris (pH 7.5) ▼
2 mM EDTA

Supplement lysis buffer with 1 mM PMSF ▼, 10 µg/ml aprotinin ▼, and 10 µg/ml leupeptin, immediately before use.

RF10 Medium PROT

RPMI 1640 media (serum free) (Invitrogen)
10% fetal calf serum
1% penicillin-streptomycin ▼
2% glutamine

▪ Chromatin Immunoprecipitation

In reference material, the following buffers are provided to map genomic targets of nuclear proteins in cultured cells.

Fixation Solution PPI2

11% formaldehyde ▼ (from a 37% stock equilibrated with methanol)
100 mM NaCl
1 mM EDTA
0.5 mM EGTA
50 mM HEPES (pH 8)
100 mM PMSF (phenylmethylsulfonyl fluoride) ▼ (in isopropanol; stable at room temperature)

Prepare fresh before use.

Cell Lysis Buffer PPI2

5 mM PIPES (pH 8)
85 mM KCl ▼
0.5% Nonidet P-40
1 mM PMSF ▼
proteinase inhibitors ▼ (leupeptin, aprotinin, pepstatin; each final concentration 2 µg/ml)

Stable at room temperature; add PMSF and proteinase inhibitors and place on ice before use.

Nuclear Lysis Buffer PPI2

50 mM Tris-HCl (pH 8)
10 mM EDTA
0.8% SDS (sodium dodecyl sulfate) ▼
1 mM PMSF ▼
protease inhibitors ▼ (leupeptin, aprotinin, pepstatin; each final concentration
  2 μg/ml)

Stable at room temperature; add PMSF and proteinase inhibitors and place
on ice before use.

Dilution Buffer PPI2

10 mM Tris-HCl (pH 8.0)
0.5 mM EGTA
1% Triton X-100 ▼
140 mM NaCl
1 mM PMSF ▼
protease inhibitors ▼ (leupeptin, aprotinin, pepstatin; each final concentration 2 μg/ml)

Stable at room temperature; add PMSF and proteinase inhibitors and place
on ice before use.

RIPA Buffer PPI2

10 mM Tris-HCl (pH 8.0)
1 mM EDTA
0.5 mM EGTA
1% Triton X-100 ▼
0.1% sodium deoxycholate ▼
0.1% SDS ▼
140 mM NaCl
1 mM PMSF ▼

Stable at room temperature; add PMSF before use.

LiCl Buffer   PPI2

0.25 M LiCl ▼
0.5% Nonidet P-40
0.5% sodium deoxycholate ▼
1 mM EDTA
10 mM Tris-HCl (pH 8.0)

Stable at room temperature.

Ligase Buffer   PPI2

12.5 mM $MgCl_2$ ▼
25 mM dithiothreitol (DTT) ▼
1.25 mM ATP
50 mM Tris-HCl (pH 7.6)

Hybridization Buffer   PPI2

7% SDS ▼
1 mM EDTA
1% bovine serum albumin (BSA)
0.5 M $NaHPO_4$ (pH 7.2) ▼ *

*This is 0.25 M Na_2HPO_4 with the pH adjusted to 7.2 with orthophosphoric
  acid.

Wash Buffer   PPI2

5% SDS ▼
1 mM EDTA
0.5% BSA
40 mM $NaHPO_4$ (pH 7.2) ▼

▪ Use/Labeling of GST Fusions

The following buffers are designed for preparing GST fusions from IPTG inducible bacterial expression vectors, GST pulldowns, and far western protocols.

2x SDS-PAGE Sample Buffer PPI2

100 mM Tris-HCl (pH 6.8)
4% (w/v) SDS ▼ (electrophoresis grade)
0.2% (w/v) bromophenol blue ▼
20% (v/v) glycerol
200 mM dithiothreitol (DTT) ▼ or β-mercaptoethanol ▼

Lysis Buffer for GST Pulldown (Ice Cold) PPI2

20 mM Tris (pH 8.0) ▼
200 mM NaCl
1 mM EDTA (pH 8.0)
0.5% Nonidet P-40
2 µg/ml aprotinin ▼
1 µg/ml leupeptin ▼
0.7 µg/ml pepstatin ▼
25 µg/ml PMSF ▼

2x PK Buffer PPI2

100 mM potassium phosphate buffer (pH 7.15) ▼
20 mM $MgCl_2$ ▼
10 mM NaF ▼
9 mM DTT ▼

Basic Buffer PPI2

20 mM HEPES (pH 7.5)
50 mM KCl ▼
10 mM $MgCl_2$ ▼
1 mM DTT ▼
0.1% Nonidet P-40

■ **Chemical Cross-linking**

The following SDS buffer is used in simple cross-linking screen using sulfo-*m*-maleimidobenzoyl-*N*-hydroxysulfo-succinimide (MBS) ester.

SDS Buffer PPI2

0.125 M Tris (pH 6.8) ▼
20% glycerol
5% β-mercaptoethanol ▼
4% SDS ▼
0.003% Coomassie blue ▼ (Sigma-Aldrich)

■ **Screening Phage Libraries/Phage Display**

The following buffers are used in overlay screening of λ-phage libraries to identify interacting proteins or engineering of protein domains.

10x Labeling Buffer PPI2

500 mM HEPES-KOH (pH 7.0)
100 mM MgCl$_2$ ▼
40 mM MnCl$_2$ ▼
10 mM DTT (dithiothreitol) ▼ (add fresh, just before use)
1 mg/ml BSA (bovine serum albumin)

Lambda Dilution Buffer PPI2

35 mM Tris-HCl (pH 7.5)
100 mM NaCl
10 mM MgCl$_2$ ▼
0.01% gelatin (Sigma-Aldrich)

Autoclave.

Blocking Solution PPI2

PBS with
    5% instant nonfat dry milk
    0.5 mM phenylmethylsulfonyl fluoride (PMSF) ▼ (Sigma-Aldrich) (add fresh)
    0.2 mM benzamidine-HCl ▼ (Sigma-Aldrich) (add fresh)

Binding Buffer PPI2

25 mM Tris-HCl (pH 7.5)
100 mM KCl ▼
5 mM MgCl₂ ▼
1 mM DTT ▼ (add fresh)
5% glycerol
1% nonfat dry milk
0.1% Tween-20
0.4 mM 4-(2-aminoethyl)benzenesulfonylfluoride (AEBSF; Calbiochem)
0.1 mM benzamidine-HCl (Sigma-Aldrich)
1 µg/ml aprotinin ▼ (Sigma-Aldrich)
1 µg/ml leupeptin ▼ (Sigma-Aldrich)
1 mM NaF ▼
0.1 mM sodium orthovanadate (Na₃VO₄) ▼

Washing Solution PPI2

25 mM Tris-HCl (pH 7.5)
100 mM KCl ▼
5 mM MgCl₂ ▼
0.1 mM DTT ▼ (add fresh)
5% glycerol
0.1% Tween-20
0.4 mM PMSF ▼
0.2 mM benzamidine-HCl (Sigma-Aldrich)
1 mM NaF ▼
0.1 mM Na₃VO₄ ▼

Blocking Buffer PPI2

3% (w/v) BSA (bovine serum albumin) diluted in PBS

SB Medium PPI2

30 g of tryptone (BD Biosciences, Difco 0123-17-3)
20 g of yeast extract (BD Biosciences, Difco 0127-17-9)
10 g of MOPS ▼ (3[*N*-morpholino] propanesulfonic acid; Sigma-Aldrich M 8899)

Bring to 1 liter total volume with dH₂O, stir until dissolved, and titrate to pH 7. Sterilize by autoclaving at 15 psi on liquid cycle for 20 min at 121°C.

■ Two-hybrid Systems

The buffers detailed below are commonly used in two-hybrid systems to identify interacting proteins.

Lysis Solution   PPI2

Zymolyase 100T ▼ dissolved 2–5 mg/ml in rescue buffer
or
β-glucuronidase ▼ , 100,000 units/ml (Sigma-Aldrich), 1:50 in Y-lysis buffer

Sterile Glycerol Solution for Freezing Transformants   PPI2

65% sterile glycerol
0.1 M $MgSO_4$ ▼
25 mM Tris-HCl (pH 8.0)
20 mg/ml X-Gal in dimethylformamide (DMF) ▼

Rescue Buffer   PPI2

50 mM Tris-HCl (pH 7.5)
10 mM EDTA
0.3% β-mercaptoethanol ▼
Prepare fresh solution for each use.

Z Buffer   PPI2

$Na_2HPO_4 \cdot 7H_2O$ ▼	16.1 g
$NaH_2PO_4 \cdot H_2O$ ▼	5.5 g
KCl ▼	0.75 g
$MgSO_4 \cdot 7H_2O$ ▼	0.246 g

Add distilled water to 1 liter and filter-sterilize through a 0.2-μm filter; store at room temperature.

Add 2.7 μl of β-mercaptoethanol ▼ per 1 ml just before use. Z buffer containing β-mercaptoethanol can be stored for approximately 1 month at 4°C.

■ Using Budding Yeast

The following buffers are used to identify protein–protein interactions using budding yeast.

Synthetic Dextrose PPI2

2% dextrose
0.17% yeast nitrogen base
0.5% ammonium sulfate ▼

YPDA PPI2

1% yeast extract
2% dextrose
2% peptone supplemented with 10 mg/liter adenine

YRLB (0.2 ml Required) PPI2

0.5 M NaCl
0.2 M Tris-HCl (pH 7.5)
10 mM EDTA
1% SDS ▼

▪ Calorimetry Analysis of Protein–Protein Interactions

The following buffers are used in investigating interactions between proteins and a low-molecular-weight ligand.

Buffer 1 PPI2

50 mM Tris-HCl (pH 8.0)
5 mM $MgCl_2$ ▼
Filter and degas.

Buffer 2 PPI2

50 mM Tris-HCl (pH 7.4)
100 mM NaCl
5 mM $MgCl_2$ ▼
Filter and degas.

▪ *X-ray Crystallography*

The following buffers are used for cocrystallization for X-ray structure determination.

Crystal Harvest Buffer   PPI2

30% PEG-MME 5000 ▼
0.1 M Bis-Tris (pH 6.5)

Protein Lysate Buffer   PPI2

20 mM HEPES (pH 7.5)
50 mM NaCl
10 mM dithiothreitol (DTT) ▼
1 mM phenylmethylsulfonyl fluoride (PMSF) ▼

Protein Elution Buffer   PPI2

20 mM HEPES (pH 7.5)
500 mM NaCl
10 mM dithiothreitol (DTT) ▼
1 mM phenylmethylsulfonyl fluoride ▼

Gel Filtration Buffer   PPI2

20 mM HEPES (pH 7.5)
50 mM NaCl
10 mM dithiothreitol (DTT) ▼

▪ *Mass Spectrometry*

The following buffers are used for preparing proteins (by enzymatic digestion and silver staining), mass spectrometry, and identifing novel protein complexes.

Acetic Acid Fix and Stop Solution   PPI2

94% (v/v) H$_2$O
5% (v/v) ethanol ▼
1% (v/v) acetic acid ▼

Developing Solution *(for 100 ml)* PPI2

6 g of Na_2CO_3 ▼
50 ml of 37% formaldehyde ▼
2 ml of gel-sensitizing solution (see the following page)

Ethanol Rinse Solution PPI2

50% (v/v) H_2O
10% (v/v) ethanol ▼

Gel-sensitizing Solution PPI2

$Na_2S_2O_3 \cdot 5H_2O$ (20 mg/ml)

Digestion Buffer PPI2

50 mM NH_4HCO_3 ▼
5 mM $CaCl_2$ ▼

HPLC Separation Buffer A PPI2

95% H_2O
5% acetonitrile ▼
0.1% formic acid ▼

HPLC Separation Buffer B PPI2

20% H_2O
80% acetonitrile ▼
0.1% formic acid ▼

HPLC Separation Buffer C PPI2

95% H_2O
5% acetonitrile ▼
0.1 formic acid ▼
500 mM ammonium acetate

▪ Identifying Novel Protein Kinase Substrates

The following buffers are used for screening and identifying novel kinase substrates.

M2 Lysis Buffer (Cold) PPI2

50 mM Tris-base (pH 7.4)
150 mM NaCl
10% glycerol
1% Triton X-100 ▼
0.5 mM EDTA
0.5 mM EGTA
50 mM NaF ▼
40 mM β-glycerophosphate
5 mM tetrasodium pyrophosphate ▼
0.1 mM sodium vanadate ▼
10 μg/ml aprotinin ▼
5 μg/ml leupeptin ▼
2 mM phenylmethylsulfonyl fluoride (PMSF), fresh ▼

10x Kinase Buffer PPI2

250 mM HEPES (pH 7.4)
100 mM magnesium acetate ▼
10 mM dithiothreitol (DTT) ▼

Kinase Reaction Buffer A (for Analog Inhibition Assay) PPI2

1x kinase buffer
10 μM ATP
100 μM ATP analog
10 μCi/reaction [γ-^{32}P]ATP ▼
1–5 μg of a known substrate/reaction

Kinase Reaction Buffer B (for Substrate Phosphorylation Assay) PPI2

1x kinase buffer
100 μM ATP analog
1–5 μg of a known substrate/reaction
10 μCi/reaction [γ-^{32}P]ATP analog ▼ (required only if a phospho-specific antibody to the substrate is not available)

Ponceau Solution PPI2

0.5% (w/v) in 1% acetic acid ▼

2x Laemmli Sample Buffer PPI2

100 mM Tris (pH 6.8) ▼
2% SDS ▼
20% glycerol
4% β-mercaptoethanol ▼ (added fresh)

Hypotonic Lysis Buffer PPI2

20 mM HEPES (pH 7.4)
2 mM EGTA
2 mM MgCl₂ ▼

Two-dimensional Sample Buffer PPI2

7 M urea ▼
2 M thiourea ▼
4% 3-[(3-Cholamidopropyl)dimethyl-ammonio]-1-propanesulfonate (CHAPS)
2% dithiothreitol (DTT) ▼

SDS-equilibration Buffer PPI2

50 mM Tris-HCl (pH 8.8)
6 M urea ▼
30% (v/v) glycerol
2% (w/v) SDS ▼
bromophenol blue for color (~0.002% w/v) ▼
10 mg/ml dithiothreitol ▼ (added just before use)

1% Agarose in SDS Running Buffer PPI2

1% agarose dissolved in SDS running buffer
    14.4 g of glycine
    3 g of Tris base
    1 g of SDS ▼ in 1 liter of H₂O
Add bromophenol blue ▼ to obtain a dark blue color (~0.002% w/v).

▪ Ribosome Display

The following buffers are used in the process of preparing protein-tethered ribosomes.

PremixZ PPI2

250 mM Tris-acetate (pH 7.5 at 4°C)
1.75 mM of each amino acid, except methionine
10 mM ATP
2.5 mM GTP
5 mM cAMP
150 mM acetylphosphate
2.5 mg/ml *E. coli* tRNA
0.1 mg/ml folinic acid
7.5% PEG-8000 (Sigma-Aldrich)

Washing Buffer PPI2

50 mM Tris-acetate (pH 7.5 at 4°C)
150 mM NaCl
50 mM magnesium acetate ▼
0.1% Tween-20
2.5 mg/ml heparin (Sigma-Aldrich)
RNase-free H_2O (DEPC-treated)

▪ Peptide Arrays

The following buffers and solutions are used in the preparation and synthesis of membrane-bound peptide arrays.

Citrate-buffered Saline PPI2

8.0 g of NaCl
0.2 g of KCl ▼
10.51 g of citric acid ▼ (x1 H_2O) in 1 liter of H_2O
Adjust pH to 7.0 with NaOH ▼; autoclave and store at 4°C.

Color Developing Solution (CDS) PPI2

Dissolve 50 mg of (3-[4,5-dimethylthiazol-2-yl]-2,5-diphenyltetrazolium bromide (MTT) in 1 ml of 70% DMF ▼ /30% water; store at –20°C. Dissolve 60 mg of 5-bromo-4-chloro-3-indolylphosphate *p*-toluidine salt (BCIP) ▼ in 1 ml of DMF ▼ ; store at –20°C. Always prepare fresh CDS on the day of the experiment: To 10 ml of CBS, add 50 µl of 1 M MgCl$_2$ ▼ , 40 µl of BCIP, and 60 µl of MTT.

Never use NBT (4-nitro blue tetrazolium chloride) instead of MTT, because the developed color will not be removable from the membrane.

Membrane Blocking Solution PPI2

Mix 20 ml of casein-based blocking buffer concentrate (Sigma-Aldrich-Genosys SU-07-250), 80 ml of T-TBS (pH 8.0), and 5 g of sucrose. The resulting pH will be 7.0; store at 4°C.

Transfer Buffer for Western Blotting PPI2

25 mM Tris-HCl (pH 7.6)
192 mM glycine ▼
20% methanol ▼
0.03% sodium dodecyl sulfate (SDS) ▼

Deprotection Mix PPI2

trifluoroacetic acid (TFA, synthesis grade) ▼
triisobutylsilane ▼ (TIBS)
H$_2$O
dichloromethane (DCM) ▼
Prepare in a ratio of 80% TFA, 3% TIBS, 5% H$_2$O, and 12% DCM.

Stripping Mix A PPI2

8 M urea ▼
1% SDS ▼ in PBS

Store at room temperature. Add 0.5% 2-mercaptoethanol ▼ before use and adjust pH to 7.0 with acetic acid ▼ .

Stripping Mix B PPI2

10% acetic acid ▼
50% ethanol ▼
40% H₂O
Store at room temperature.

Acetylation Mix PPI2

22% solution of acetic anhydride (≥ 99.5%) ▼ in DMF ▼

■ Fluorescence Resonance Energy Transfer (FRET) Analysis

The following buffers and solutions are used in the preparation of cell extracts for visualization of labeled proteins (see also Section 3, Imaging).

Paraformaldehyde Fixative Solution PPI2

Dissolve 4 g of paraformaldehyde ▼ (Sigma-Aldrich) in 50 ml of distilled, deionized H₂O, and then add 1 ml of 1 M NaOH ▼ solution. Stir gently on a heating block (~65°C) until the paraformaldehyde is dissolved. Add 10 ml of 10x PBS and allow to cool to room temperature. Adjust the pH to 7.4 with 1 M HCl ▼ (~1 ml), then adjust the final volume to 100 ml with distilled, deionized H₂O. Filter the solution through a 0.45-μm membrane filter to remove any particulate matter, and store in aliquots for several months at −20°C. Avoid repeated freeze/thawing of the paraformaldehyde solution.

Quenching Solution PPI2

50 mM Tris-HCl (pH 8.0)
100 mM NaCl

Gelatin Solution PPI2

Dissolve 0.5 g of gelatin (porcine, Sigma-Aldrich) in 500 ml of 1x PBS. Sterilize by autoclaving.

Mowiol Mounting Solution PPI2

This is the desired mounting medium, because it does not quench GFP fluorescence. Mix 6 ml of glycerol, 2.4 g of Mowiol 4-88 (Calbiochem), and 6 ml of distilled, deionized H_2O. Initially shake for 2 hours at room temperature, then add 12 ml of 200 mM Tris-HCl (pH 8.5) and incubate at 50°C with occasional mixing until the Mowiol dissolves (~3 hr). Filter through a 0.45-μm membrane filter and store in aliquots for weeks at 4°C, or for months at −20°C.

▪ *Protein Fragment Complementation Assays*

The following buffers and reagents are used in protein fragmentation assays and screening of degenerate libraries.

DHFR PCA Selective Medium PPI2

For 1 liter of solid agar-M9 minimal medium
    777 ml of 2.5% noble agar (Difco)
    200 ml of 5× M9 salts solution
    20 ml of 20% glucose solution
    2 ml of 1 M $MgSO_4$ ▼
    1 ml 100 mM $CaCl_2$ ▼
Supplement the medium with the following antibiotics and nutrients
    100 μg/ml ampicillin ▼
    25 μg/ml kanamycin ▼
    1 mM isopropyl-β-D-thio-galacto-pyranoside (IPTG) ▼
    10 μg/ml trimethoprim ▼
    800 μg/ml casamino acids (Difco)
    10 μg/ml thiamine

All solutions must be prepared with deionized H_2O and filter-sterilized in the case of antibiotics, casamino acids, IPTG, and thiamine (store at −20°C), and sterilized by autoclaving in the case of salts (store at room temperature). Pour the reconstituted medium into 100-mm or 150-mm petri dishes. Store medium for up to 2 months at 4°C.

Casamino acids and thiamine are added because they help *E. coli* cells to grow under DHFR PCA selection without loss of stringency. Alternatively, thymine can be used.

Renilla *Luciferase Assay Buffer*   PPI2

50 mM KCl ▼
50 mM NaCl
20 mM HEPES (pH 8)
5 mM DTT ▼
2.5 mM MgCl$_2$ ▼
2 mM EDTA
0.2% Nonidet P-40 (w/v)

Lysis Buffer   PPI2

50 mM Tris-HCl (pH 7.6)
150 mM NaCl
2% SDS ▼
5 mM EDTA
100 μg/ml PMSF ▼

■ *Electron Microscopy*

The following solutions are used for preparing tissues to visualize proteins by electron microscopy.

Epon Embedding Medium ▼   PPI2

Made as a stock from
     86.0 g of epoxy embedding medium
     100 g of Araldite M Hardener 964
     28 g of Epon Hardener MNA
     3 ml of Epoxy Embedding Medium
     accelerator
All ingredients are from Fluka.

Substitution Media   PPI2

For analysis of morphology
     0.25% glutaraldehyde ▼
     0.5% osmium tetroxide ▼
     dry acetone ▼
For immunolabeling of specific proteins
     0.1% uranyl acetate ▼
     0.1% glutaraldehyde ▼
     dry acetone ▼

Label Buffer PPI2

0.1% (w/v) BSA in PBS (pH 7.4) (Aurion)

▪ Protein Complex Characterization

The following solutions are used for dissecting protein–protein interactions using affinity chromatography.

Immobilization Buffer PROT

phosphate-buffered saline (PBS)
0.02% (v/v) Tween-20
0.5 mM Tris(2-carboxyethyl)-phosphine hydrochloride (TCEP) ▼ (Pierce)

Lysis Buffer PROT

10 mM Tris-Cl (pH 7.5)
100 mM NaCl
0.5% Nonidet P-40

Immediately before use, add the protease inhibitor 1 mM phenylmethylsulfonyl fluoride (PMSF) ▼ in addition to Complete Protease Inhibitor Cocktail tablets (Roche) and the protein phosphatase inhibitors 1 mM sodium orthovandate (Na_3VO_4) ▼ and 1 mM sodium fluoride (NaF) ▼ .

▪ Tandem Affinity Chromatography

IPP150 Calmodulin-binding Buffer PPI2

10 mM β-mercaptoethanol ▼ (add immediately before use)
10 mM Tris-HCl (pH 8.0)
150 mM NaCl
1 mM magnesium acetate ▼
1 mM imidazole ▼
2 mM $CaCl_2$ ▼
0.1% IGEPAL CA-630 (Nonidet P-40–type nonionic detergent; Sigma-Aldrich I-8896)

IPP150 Calmodulin Elution Buffer PPI2

10 mM β-mercaptoethanol ▼ (dilute immediately before use)
10 mM Tris-HCl (pH 8.0)
150 mM NaCl
1 mM magnesium acetate ▼
1 mM imidazole ▼
2 mM EGTA
0.1% IGEPAL CA-630

IPP150 Buffer PPI2

10 mM Tris-HCl (pH 8.0)
150 mM NaCl
0.1% IGEPAL CA-630

TEV Cleavage Buffer PPI2

10 mM Tris-HCl (pH 8.0)
150 mM NaCl
0.1% IGEPAL CA-630
0.5 mM EDTA
1 mM DTT ▼ (add immediately before use from 1 M stock)

■ *Nuclear Magnetic Resonance (NMR) Imaging*

The following protocols are used for high-throughput screening of protein–protein interactions using NMR.

MDM2 Buffer PPI2

50 mM MES-d13 (pH 7.0) (Cambridge Isotopes)
150 mM KCl ▼
50 mM DTT-d1 ▼ (Cambridge Isotopes)
1.5 mM NaN$_3$ ▼

HDM2 Buffer I PPI2

100 mM potassium phosphate (pH 7) ▼
1 mM EDTA
2 mM DTT ▼

HDM2 Buffer 2   PPI2

20 mM HEPES (pH 7)
50 mM NaCl
5 mM DTT▼

General Information for Proteomics

Commonly Used Buffers for Cation-exchange Chromatography   PUR

Substance	pK_a	Working pH
Citric acid ▼	3.1	2.6–3.6
Lactic acid	3.8	3.4–4.3
Acetic acid ▼	4.74	4.3–5.2
MES[a] ▼	6.1	5.6–6.6
ADA[b]	6.6	6.1–7.1
MOPS[c] ▼	7.2	6.7–7.7
Phosphate	7.2	6.8–7.6
HEPES[d]	7.5	7.0–8.0
Bicine[e]	8.3	7.6–9.0

[a]2-[N-morpholino]ethanesulfonic acid.
[b]N-[2-acetamido]-2-iminodiacetic acid.
[c]3-[N-morpholino]propanesulfonic acid.
[d]N-[2-hydroxyethyl]piperazine-N'-[2-ethanesulfonic acid].
[e]N,N-bis[2-hydroxyethyl]glycine.

Commonly Used Buffers for Anion-exchange Chromatography   PUR

Substance	pK_a	Working pH
N-Methyl-piperazine ▼	4.75	4.25–5.25
Piperazine ▼	5.68	5.2–6.2
Bis-Tris	6.5	6.0–7.0
Bis-Tris propane	6.8	6.3–7.3
Triethanolamine ▼	7.8	7.25–8.25
Tris ▼	8.1	7.6–8.6
N-Methyl-diethanolamine ▼	8.5	8.0–9.0
Diethanolamine ▼	8.9	8.4–9.4
Ethanolamine ▼	9.5	9.0–10.0
1,3-Diaminopropane	10.5	10.0–11.0

Volatile Buffering Salts PUR

Substance	pK_a	Working pH	Counterion
Formic acid ▼	3.8	3.3–4.3	NH_3^+
Acetate	4.7	4.3–5.2	NH_4^+
Pyridine/acetate ▼	5.35/4.74	4.3–5.9	CH_3COO^-
N-Ethylmorpholine ▼	7.7	7.2–8.2	CH_3COO^-
Ammonium	9.25	8.7–9.7	CH_3COO^-
Trimethylamine ▼	9.8	9.3–10.3	CH_3COO^-

Specificity of Immunoprecipitation Conjugates PPI2

Protein	Specificity
Protein A/Staph-A cells	human IgG1, IgG2, IgG4; mouse IgG2a, IgG2b, IgA; polyclonals from rabbit, mouse, guinea pig, pig, cat, dog, monkey
Protein G	human IgG1, IgG2, IgG3, IgG4; mouse IgG1, IgG2a, IgG2b, IgG3; rat IgG1, IgG2a, IgG2b, IgG2c; polyclonal antibodies from rabbit, mouse, goat, cow, sheep, horse, donkey, monkey
Protein L	all mouse, rat, human IgG types; polyclonal antibodies from mouse, rat, hamster, pig, chicken

Liquid Chromatography Columns PPI2

Column	MW range	Type	Bed volume (ml)	pH range	Capacity (mg)	Ordering information[a]
Mono Q HR 5/5	up to 10^7	exchange anion	1	3–11	25	17-0546-01
Superdex 200 HiLoad 16/60	1×10^4– 6×10^5	gel filtration	120	3–12	60	17-1069-01
Superdex 200 HR 10/30	1×10^4– 6×10^5	gel filtration	24	3–12	5–12	17-1088-01
Superose 6 HR 10/30	5×10^3– 5×10^6	gel filtration	24	3–12	5–12	17-0537-01

[a]GE Healthcare, www.gehealthcare.com

NOTES

NOTES

SECTION 5
Arabidopsis

Arabidopsis

The following section contains buffers and reagents used in plant studies (specifically *Arabidopsis*) that are not covered in previous sections. Areas include culturing conditions, as well as preparation of tissues for histochemistry or isolation of nucleic acid.

Caution: See Cautions Appendix for appropriate handling of materials marked with ▼.

Microscopy

The following section contains recipes for the preparation of commonly used fixatives. Because *Arabidopsis* seedlings are small, whole seedlings (1–2 weeks) can be fixed (including the root system) without dissection. Before embedding, fixed material can be dissected to isolate organs of interest such as individual flowers. Cut the leaves from plants older than 1–2 weeks into 1-cm^2 pieces.

For Standard Microscopy

Fixative ARA

1.5% (v/v) glutaraldehyde (EM grade, 25% [v/v] stock solution) ▼
0.3% (v/v) paraformaldehyde ▼
0.025 M PIPES buffer
0.1% (v/v) Tween-20 (not required when fixing roots)

For Transmission Electron Microscopy

TEM Fixative ARA

4% (v/v) paraformaldehyde ▼
0.1% (v/v) glutaraldehyde ▼
0.05% (v/v) Tween-20 in 100 mM PIPES (pH 7.1)

Genetic Manipulation

The following section contains recipes for the preparation of commonly used buffers in root transformation.

Root Transformation

Hormone Stocks ARA

1000x N^6-(2-isopentenyl)adenine (2-iP)   5 mg/ml
2000x 2,4-D   1 mg/ml
2000x indole-3-acetic acid (IAA)   0.2 mg/ml
2000x kinetin   0.1 mg/ml
1000x α-naphthylphthalamic acid (NAA) 2 mg/ml

Prepare hormone stocks in high-quality DMSO and store aliquots at –20°C.

Germination Medium (for 1 Liter) ARA

1x MS salts
10 g of sucrose
0.5 g of MES ▼

Adjust pH to 5.7 with 1 N KOH ▼ and add 7.5 g of tissue-culture-grade agar. After autoclaving, cool medium to ~60°C and then add 1 ml of 1000x Gamborg's B5 vitamins.

Callus-inducing Medium ARA

1x Gamborg's B5 salts
20 g of glucose
0.5 g of MES ▼

Adjust pH to 5.7 with 1 N KOH ▼.

For solid medium, add 7.5 g of tissue-culture-grade agar.

After autoclaving, cool medium to approximately 60°C and then add:

    500 μl of 2000x 2,4-D (final concentration 0.5 mg/liter)
    500 μl of 2000x kinetin (final concentration 0.05 mg/liter)

Shoot-inducing Medium ARA

1x Gamborg's B5 salts
20 g of sucrose
0.5 g of MES ▼

Adjust pH to 5.7 with 1 N KOH ▼ and add 7.5 g of tissue-culture-grade agar. After autoclaving, cool medium to approximately 60°C and then add:

  1 ml of 1000x 2-iP (final concentration 5 mg/liter)
  500 µl of 2000x IAA (final concentration 0.1 mg/liter)
  50 mg/liter kanamycin (for transgene selection)
  500 mg/liter timentin or carbenicillin (to eliminate *Agrobacterium* after
    transformation)

Prepare fresh by dissolving 500 mg of timentin or carbenicillin in 10 ml of double-distilled water and filter-sterilize.

Protein Extraction

The homogenization buffer detailed is used in extraction of protein from *Arabidopsis*; however, it is often also preferable to prepare subcellular fractions that allow localization of protein expression and improve sensitivity. Buffers used in organelle preparation are also provided.

Homogenization Buffer ARA

50 mM Tris acetate (pH 7.9)
100 mM potassium acetate
1 mM EDTA
1 mM DDT (dithiothreitol) ▼
20% glycerol
protease inhibitors of choice

Xpl Homogenization Buffer ARA

sorbitol	30.1 g
1 M HEPES (pH 7.5)	25 ml
0.5 M EDTA	2 ml
1 M MgCl₂ ▼	0.5 ml
distilled H₂O	to 500 ml

After autoclaving, allow the solution to cool and then add 1.25 g of BSA and 0.5 g of sodium ascorbate.

PBF Percoll (for Two Gradients) ARA

PEG 4000 (polyethylene glycol) ▼	450 mg
BSA	150 mg
Ficoll	150 mg
Percoll	15 ml

Prepare the previous day and store overnight at 4°C.

Nuclear Extraction

Grinding Buffer ARA

10% glycerol
400 mM sucrose
50 mM Tris (pH 8.0) ▼
H_2O

Just before use, add 1 μl of 100 mM PMSF and keep the buffer aliquots on ice.

Histochemistry of GUS Fusion

The following GUS staining solution is used in staining proteins targeted to the nucleus.

GUS Staining Solution ARA

100 mM sodium phosphate (pH 7.0) ▼
1 mM EDTA (pH 8)
1% Triton X-100 ▼
5 mM potassium ferrocyanide ▼
5 mM potassium ferricyanide ▼
1 mg/ml X-Gluc (freshly predissolved in a drop of *N,N*-dimethyl-
  formamide ▼)

Staining solution can be stored at −20°C and reused several times.

NOTES

NOTES

Murine

Murine

The following section contains buffers and reagents used in mouse studies that are not covered in previous sections. Areas covered include in vitro culturing of embryonic stem (ES) cells and preparation of tissues for cryostorage.

Caution: See Cautions Appendix for appropriate handling of materials marked with ▼.

In Vitro Culturing of Embryonic Stem Cells

The media below are used for in vitro culturing of ES cells.

M16 Culture Medium MME3

Compound	mM	Molecular weight	g/liter
NaCl	94.66	58.450	5.533[a]
KCl ▼	4.78	74.557	0.356
CaCl$_2$·2H$_2$O ▼	1.71	147.200	0.252
KH$_2$PO$_4$ ▼	1.19	136.091	0.162
MgSO$_4$·7H$_2$O ▼	1.19	246.500	0.293
NaHCO$_3$	25.00	84.020	2.101
Sodium lactate	23.28	112.100	2.610
			or 4.349 g
			of 60% syrup
Sodium pyruvate	0.33	110.000	0.036
Glucose	5.56	179.860	1.000
BSA			4.000
Penicillin G·potassium salt (final conc., 100 units/ml)			0.060
Streptomycin sulfate ▼ (final conc., 50 mg/ml)			0.050
Phenol red ▼			0.010[b]
2x glass-distilled H$_2$O			up to 1 liter

[a] Increase NaCl to 5.68 g/liter when CaCl$_2$ is omitted for Ca^{++}-free medium.
[b] The concentration of phenol red can be decreased to 0.0001–0.001 g/liter, because it may be embryotoxic.

M2 Culture Medium MME3

Compound	mM	Molecular weight	g/liter
NaCl	94.66	58.450	5.533[a]
KCl▼	4.78	74.557	0.356
$CaCl_2 \cdot 2H_2O$▼	1.71	147.200	0.252
KH_2PO_4▼	1.19	136.091	0.162
$MgSO_4 \cdot 7H_2O$▼	1.19	246.500	0.293
$NaHCO_3$	4.15	84.020	0.349
HEPES	20.85	238.300	4.969
Sodium lactate	23.28	112.100	2.610
			or 4.349 g
			of 60% syrup
Sodium pyruvate	0.33	110.000	0.036
Glucose	5.56	179.860	1.000
BSA			4.000
Penicillin G·potassium salt			0.060
Streptomycin sulfate▼			0.050
Phenol red▼			0.010[c]
2x glass-distilled H_2O[b]			up to 1 liter

[a] Increase NaCl to 5.68 g/liter when $CaCl_2$ is omitted for Ca^{++}-free medium.

[b] It is extremely important that the water used for making culture media be distilled at least twice in a still, in which the metal element is enclosed by glass (2x glass-distilled), or be purified by filtration (e.g., Milli-Q system producing 18 MΩ water), after an initial reverse osmosis process, and then stored in clean plastic containers. Water purified by filtration should be tested for endotoxin. Prolonged storage is not advised. It is also possible to use commercially available deionized water for tissue culture, embryo-tested or endotoxin-tested water, or water for human intravenous administration.

[c] The concentration of phenol red can be decreased to 0.0001–0.001 g/liter, because it may be embryotoxic.

KSOM Medium MME3

Component	Final concentration mм	Final concentration g/liter	Concentrated stock	Stock volume for 100 ml
			A′ (10x) g/100 ml	10 ml
NaCl	95.0	5.55	5.55	
KCl▼	2.50	0.186	0.186	
KH₂PO₄	0.35	0.0476	0.0476	
MgSO₄·7H₂O▼	0.20	0.0493	0.0493	
Glucose	0.20	0.036	0.036	
Penicillin G			0.060	0.060
Streptomycin▼			0.050	0.050
Sodium lactate	10.0	1.12 or 1.87 g of 60% syrup	1.12 or 1.87 g of 60% syrup	
			B′ (10x) g/100 ml	10 ml
NaHCO₃	25.0	2.10	2.10	
Phenol red▼			0.001	
			C′ (100x) g/10 ml	1 ml
Na pyruvate	0.20	0.022	0.022	
			D (100x) g/10 ml	1 ml
CaCl₂·2H₂O▼	1.71	0.25	0.25	
			F (1000x) g/10 ml	0.1 ml
EDTA	0.01	0.0038	0.038	
L-Glutamine	1.00	0.146	G (200x) 200 mм	0.5 ml
BSA (e.g., Sigma-Aldrich A3311)		1.000		100 mg

Isolation and Culturing of Blastocyst-derived ES Lines (e.g., Trophoblast Stem Cell Lines)

FGF4 Stock Solution (1000x, 25 µg/ml) MME3

Human recombinant fibroblast growth factor-4 (FGF4; 25 µg; Sigma-Aldrich F2278). Resuspend lyophilized FGF4 in its vial with 1.0 ml of phosphate-buffered saline (PBS)/0.1% (w/v) bovine serum albumin (BSA). Mix well, aliquot 100 µl, and freeze at −80°C. Thaw each aliquot as needed and store at 4°C. Do not refreeze.

Heparin Stock Solution (1000x, 1.0 mg/ml) MME3

Sigma-Aldrich H3149, 10,000 units. Resuspend in PBS and store at –80°C. The stock can be also prepared as a 10,000x (10 mg/ml) solution and used to make batches of 1000x.

TS Medium (650 ml) MME3

Prepare by adding the following reagents to 500 ml of RPMI medium 1640 (e.g., Invitrogen 61870 or 11875):

FBS (e.g., CanSera, CS-COB-500, may be substituted by other suppliers); 130 ml (20% final)

penicillin and streptomycin▼ (100x stock; e.g., Invitrogen 15070-063) (50 μg/ml each final)

sodium pyruvate (100 mM stock, e.g., Invitrogen 11360), 6.5 ml (1 mM final)

β-mercaptoethanol▼ (10 mM stock; e.g., Sigma-Aldrich M7522), 6.5 ml (100 μM final)

L-glutamine (200 mM stock, e.g., Invitrogen 25030 or 35050), 6.5 ml (2 mM final)

For Mitomycin C Treatment MME3

Mitomycin C▼ stock (0.5–1.0 mg/ml in PBS or glass-distilled water)

Store for 1 week at 4°C protected from light. If stored frozen, check that it is all in solution after thawing (Sigma-Aldrich M0503). Use at 10 μg/ml.

Feeder-conditioned Medium (Feeder-CM) MME3

Prepare mitomycin-treated mouse embryonic fibroblasts (MEFs) and culture in trophoblast stem (TS) medium for 72 hours. Collect the medium. Spin to remove the floating cells and debris, filter (0.45 μm), and store at –20°C in aliquots. Thaw each aliquot as needed and store at 4°C; do not refreeze. Use the MEFs to prepare two more batches of feeder-CM, then discard the cells. MEFs are used up to 10 days after mitomycin treatment.

70cond Medium MME3

Add 3 ml of TS medium into 7 ml of feeder-CM.

70cond Medium + F4H Medium   MME3

Add 10 µl of 1000x FGF4 (25 ng/ml) and 10 µl of 1000x heparin (1 µg/ml) into 10 ml of 70cond medium.

Cyropreservation

The following solutions are used to freeze mouse tissues for storage.

ES Cells

2x Freezing Medium   MME3

60% ES-DMEM (containing 15% FBS)
20% fetal bovine serum
20% dimethyl sulfoxide▼

Prepare fresh and cool on ice.

Sperm

Cryoprotective Agent (CPA)   MME3

The CPA is 18% raffinose, 3% skim milk in water.

Raffinose (Sigma-Aldrich R0250 or R7630)	1.8 g
Skim milk (Difco [B-D] Betalab 0032-17-3)	0.3 g
Ultrapure water (e.g., Sigma-Aldrich W1503)	to 10 ml
Osmolarity, 400–410 mOsm	

Genotyping

ES Cells

The following lysis buffer is used in the rapid preparation of ES cells for genotyping.

Lysis Buffer   MME3

1x PCR buffer
1.7 µM SDS▼
50 µg/ml proteinase K

1x PCR Buffer

16.6 mM ammonium sulfate▼
67 mM Tris (pH 8.8)▼
6.7 mM MgCl₂▼
6 mM β-mercaptoethanol▼
6.7 μM EDTA

Isolation of Mouse Genomic DNA from Mouse Tails

The following reagent is used in the preparation of mouse genomic DNA from tails for PCR analysis.

Digestion Buffer MME3

100 mM NaCl
50 mM Tris-HCl (pH 8)
100 mM EDTA (pH 8)
1% sodium dodecyl sulfate (SDS)▼

Immunohistochemistry of Whole-mount Embryos

The following solutions are used in the preparation and examination of whole-mount embryos.

BABB (benzyl alcohol▼ :benzyl benzoate▼ ; 1:2) MME3

BABB is used to clear the embryos after staining and to provide a mountant to observe and photograph the stained embryos.

DAB▼ (3-3′-diaminobenzidine tetrahydrochloride; Sigma-Aldrich D5637) MME3

Store desiccated at −20°C. Warm to room temperature before weighing.

DAB/NiCl₂▼ MME3

Mix 0.03 g of DAB and 0.03 g of NiCl₂ in 50 ml of PBT. Use within 1 hour of preparing and keep protected from light. The nickel enhances the sensitivity of the color reaction and produces a slate-gray–to–purple precipitate. If necessary, vary the amount of nickel to alter the intensity of the color. Cobalt can be substituted for nickel.

PBT MME3

PBS
0.2% bovine serum albumin (BSA) (Sigma-Aldrich A4378)
0.5% Triton X-100▼

Make fresh before use. (*Note:* This is different from the PBT used in whole-mount in situ hybridization.)

PBSMT MME3

PBS
2% nonfat instant skim milk (used to block all nonspecific protein-binding sites in the tissues). The brand is important. Carnation gives consistently good results, whereas other brands (e.g., Kroger) do not.
0.5% Triton X-100 ▼ (used to facilitate permeability of the tissues)

Make fresh before use.

Detergent Rinse MME3

0.1 M phosphate buffer (pH 7.3)
2 mM MgCl$_2$▼
0.01% sodium deoxycholate ▼
0.02% Nonidet P-40 (NP-40; Sigma-Aldrich N6507)

Fixative for Frozen Sections MME3

0.2% paraformaldehyde▼ in 0.1 M PIPES buffer (pH 6.9) (Sigma-Aldrich P9291)
2 mM MgCl$_2$▼
5 mM EGTA

Staining Solution MME3

0.1 M phosphate buffer (pH 7.3)
2 mM MgCl$_2$▼
0.01% sodium deoxycholate▼
0.02% Nonidet P-40
5 mM potassium ferricyanide▼
5 mM potassium ferrocyanide▼

Dilute X-gal stock to give a final concentration of 1 mg/ml. If staining is to be performed for more than 1 hour, it is best to add Tris (pH 7.3) to the staining solution to give a final concentration of 20 mM. After use, the stain can be filtered and reused and will stay fresh for months. Store at 4°C protected from light.

Other Buffers and Solutions

The following section contains general buffers and solutions commonly used in mouse studies and the examination of mouse embryos.

Acidic Tyrode Solution for Removing Zonae   MME3

	g/100 ml
NaCl	0.800
KCl▼	0.020
$CaCl_2 \cdot 2H_2O$▼	0.024
$MgCl_2 \cdot 6H_2O$▼	0.010
glucose	0.100
polyvinylpyrrolidone (PVP)▼	0.400

Prepare at room temperature and adjust to pH 2.5 with Analar HCl (BDH). The PVP is added to increase viscosity and reduce embryo stickiness. Filter-sterilize and store in aliquots at –20°C. Acidic Tyrode's solution is also available from Sigma-Aldrich (T1788) and Specialty Media (MR-004D)

Alkaline Phosphatase Buffers   MME3

For immunohistochemistry of embryo sections:
  100 mM Tris-HCl (pH 9.5)
  100 mM NaCl
  5 mM $MgCl_2$▼

For in situ hybridization of intact embryos with RNA probes (NTMT without Tween 20):
  100 mM Tris (pH 9.5)▼
  100 mM NaCl
  50 mM $MgCl_2$▼

This buffer is best if made fresh from stock solutions because precipitate tends to form on standing.

For in situ hybridization of intact embryos with RNA probes (NTMT):
  100 mM NaCl
  100 mM Tris (pH 9.5)▼
  50 mM $MgCl_2$▼
  0.1% Tween 20 (Fisher Scientific BP337-100 or Sigma-Aldrich P1379)

Buffer is best if made fresh from stock solutions because precipitate tends to form on standing.

Alkaline Phosphatase Staining Solution MME3

For culture of primordial germ cells from embryos of different stages:

Solution	Amount	Final concentration
1 M Tris-maleate (pH 9.0) (Sigma-Aldrich T3128)	1.25 ml	25 mM
α-naphthyl phosphate▼ (Sigma-Aldrich N7255)	20 mg	0.4 mg/ml
Fast-Red TR salt ▼ (Sigma-Aldrich 20,155-3)	50 mg	1 mg/ml
1 M MgCl$_2$▼	400 µl	8 mM
2x glass-distilled or Millipore Q H$_2$O		to 50 ml

Adjust pH of the Tris-maleate by mixing 1 M Tris-maleate and 1 M Tris base (Fisher Scientific BP152-1).

Ca^{++}/Mg^{++}-free Tyrode Ringer's Saline (pH 7.6–7.7) MME3

	g/liter
NaCl	8.0
KCl▼	0.3
NaH$_2$PO$_4$·5H$_2$O▼	0.093
KH$_2$PO$_4$▼	0.025
NaHCO$_3$	1.0
glucose	2.0

Chicago Sky Blue▼ (Also Called Pontamine Sky Blue; Sigma-Aldrich C8679) MME3

Dissolve in isotonic saline to provide a 1% solution (1 g/100 ml saline). Filter the solution through Whatman filter paper. Store in a glass bottle at room temperature.

Hyaluronidase MME3

Use Type IV-S from bovine testes (e.g., Sigma-Aldrich H3884 or embryo-tested H4272). Prepare a stock solution at 10 mg/ml in H$_2$O, M2, or any other HEPES-buffered embryo culture medium. Filter-sterilize, aliquot, and store at −20°C for months. Dilute to approximately 300 µg/ml in M2 or other HEPES-buffered embryo culture medium with BSA for removing cumulus cells. A 0.5–1-mg/ml final concentration of hyaluronidase also may be used; more concentrated stock solutions (e.g., 100x) are made in this case.

Mannitol (3 M) (Sigma-Aldrich M4125) MME3

Dissolve in ultrapure water, add 0.3% BSA (Sigma-Aldrich A3311), and filter through a 0.22-μm Millipore filter. Store in aliquots at −20°C. Use a fresh aliquot for each experiment.

Methylene Blue Solution MME3

Dissolve methylene blue▼ (Basic Blue 9) powder (Sigma-Aldrich M9140) in 0.5 M sodium acetate (pH 5.2)▼, to a final concentration of 0.1% w/v. Make sure to dissolve the powder completely by stirring for 30 minutes. Stain agarose gels for 20 minutes. Destain the gel in distilled H_2O for 30 minutes, changing the water every 10 minutes.

Pancreatin/Trypsin Solution for Separating Germ and Tissue Layers MME3

	g/20 ml	Final concentration
pancreatin	0.50	2.5%
trypsin	0.10	0.5%
polyvinylpyrrolidone▼ (*optional*)	0.10	0.5%

Make in Ca^{++}/Mg^{++}-free Tyrode Ringer's saline. The suspension will be difficult to filter-sterilize through a 0.45-μm Millipore filter without low-speed centrifugation or prefiltering through a Whatman No. 1 filter. Store sterile in small aliquots at −20°C.

Pronase Solution MME3

Use protease from *Streptomyces griseus* (Calbiochem 537088; Boehringer Mannheim 165 921; Sigma-Aldrich P5147). Prepare a 0.5% solution in M2 medium for removing zonae as an alternative to Acidic Tyrode's solution. If necessary, 0.5% polyvinylpyrrolidone▼ can be added to reduce stickiness of the embryos. Because Pronase is a crude enzyme preparation, it probably should be incubated for 30 minutes at room temperature to destroy contaminating nucleases, etc. Centrifuge to remove insoluble material, filter-sterilize, and store in aliquots at −20°C.

Saline/EDTA Buffer Plus Glucose for Isolation of Germ Cells and Tissue Culture   MME3

	g/100 ml
EDTA (disodium salt)	0.02
NaCl	0.80
KCl▼	0.02
Na_2HPO_4 (anhydrous)▼	0.115
KH_2PO_4▼	0.02
phenol red▼	0.001
glucose	0.02

The final EDTA concentration is 0.02%. Check that pH is 7.2. Filter-sterilize or autoclave (for 15 minutes at 121°C, 15 psi). Store at room temperature. The glucose can be omitted for tissue culture alone.

0.25% Trypsin in Tris-Saline for Tissue Culture   MME3

Trypsin stock solution for tissue culture 0.25% trypsin 1:250 in Tris-saline (or any other well-buffered isotonic salt solution). Addition of antibiotics is optional.

	g/100 ml
NaCl	8.00
KCl▼	0.40
Na_2HPO_4▼	0.10
glucose	1.00
Trizma base (Fisher Scientific BP152-1)	3.00
phenol red▼	0.010
penicillin G	0.060 (final conc. 100 units/ml)
streptomycin▼	0.100
trypsin (1:250, Difco 0152)	2.5 g
(dissolve in small volume of H_2O before adding)	

Adjust pH to approximately 7.6. Filter-sterilize and aliquot into sterile containers. Store at −20°C. Dilute stock 1:4 in saline/EDTA buffer before use.

NOTES

NOTES

SECTION 7
Storage and Shipment of Biological Samples

Storage and Shipment of Biological Samples

This section provides commonly used methods for storing and shipping various types of biological samples: bacterial stocks, yeast stocks, mammalian tissue-culture cells, blood or tissue samples, and DNA. All methods are updated from B. Birren, E.D. Green, R.M. Myers, and J. Roskins, eds. 1998. *Genome analysis: A laboratory manual.* Cold Spring Harbor Laboratory Press, Cold Spring Harbor, New York. Compliance with local, state, and federal regulations for the shipment of biohazardous materials is the responsibility of the investigator. Consult your local institution for further guidelines.

> *Caution:* See Cautions Appendix for appropriate handling of materials marked with ▼.

Bacterial Stocks

Most bacterial stocks are frozen in 7% DMSO ▼ or 15% glycerol at –80°C for long-term storage. Viability of frozen cells depends on the specific strain and the health of the cells at the time of freezing. Cultures to be stored are typically started from a single colony and grown in a suitable medium with agitation overnight (~10–15 hours).

DMSO Stocks

Transfer 1 ml of an overnight culture into a labeled 1.5-ml screwcap cryotube and add 80 μl of DMSO ▼. (Use DMSO from a bottle specifically dedicated for bacterial stock preparation. Never pipette directly from the stock bottle; aseptically remove an aliquot from the bottle and use the aliquot of DMSO to prepare the cultures.) Cap the tube and mix gently. Store at –80°C. Long-term viability of stocks depends on the particular strain, but some bacterial stocks have been known to maintain good viability for up to 10 years after initial storage in DMSO.

Glycerol Stocks

Transfer 0.5 ml of an overnight culture into a labeled 1.5-ml screwcap cryotube and add 0.5 ml of sterile 30% glycerol. Cap the tube and mix gently. Store at –80°C. Long-term viability of stocks depends on the particular strain. Alternatively, grow bacteria in medium containing 8–10% glycerol in plastic multiwell plates and store at –80°C. This method is typically used for storing cosmid, bacteriophage P1, BAC, and cDNA libraries.

Retrieval of Frozen Bacterial Stocks

Never thaw frozen bacterial stocks in DMSO or glycerol. Use a sterile loop, sterile wooden stick, or sterile disposable pipette to scratch the surface of the stock. Streak appropriate agar plates (e.g., LB agar plates) for single colonies. Recap the frozen stock and return it to storage at –80°C. Incubate the plate overnight at 37°C. The colonies on a plate can be used for up to 1 week to inoculate cultures. Plates should be stored upside down at 4°C during this time.

Shipment of Bacterial Strains by Mail

In general, most bacterial strains can be shipped by several different methods that maintain good viability. Agar stab cultures have traditionally been used to store and ship bacterial strains. Parafilm-sealed petri dishes streaked for single colonies or sterile filter disks impregnated with bacterial culture can also be shipped. The latter are aseptically transferred to appropriate medium on receipt and a fresh overnight culture is grown. Overnight cultures can be shipped in cryotubes at room temperature. On receipt, the culture is streaked on plates of appropriate selective medium and single colonies are isolated. Frozen DMSO and glycerol stocks can be shipped on dry ice.

Agar stab cultures of nonplasmid-containing strains can be stored at room temperature for many years. They are not appropriate for long-term storage of plasmid-bearing strains because plasmid can be lost under nonselective conditions. However, plasmid-bearing strains can be conveniently shipped in agar stab cultures. Immediately on receipt, the cells must be streaked on plates of selective medium and either DMSO or glycerol stocks should be made.

Agar stab cultures are prepared in 3-ml glass vials with rubber gaskets in the screwcaps (e.g., Wheaton) as follows: Place 2 ml of liquified LB top agar (0.7% agar; some investigators use 1–1.2% agar) in each vial. Autoclave the vials with the caps loosened. Allow to cool to room temperature, tighten the caps, and store at room temperature until needed. Use a sterile loop or sterile wooden stick to pick an isolated single colony and stab it through the center of the agar to the bottom of the vial. Tighten the cap.

Bacterial strains should only be shipped at room temperature in moderate weather conditions. Hot summer weather may be lethal during shipment.

Yeast Stocks

Yeast stocks are typically frozen in 20% glycerol at –80°C for long-term storage. Viability of frozen yeast cells depends on the specific strain and the health of the cells at the time of freezing. Cultures to be stored are typically started from "patched out" clones and grown in a suitable medium with agitation overnight.

Yeast Glycerol Stocks

Patch out clones on an appropriate plate. Incubate for 2 days at 30°C. Inoculate 6 ml of YPD medium with a "matchhead-full" of cells from the plate (this is a large inoculum; the culture will be turbid before incubation). Incubate with agitation overnight at 30°C. Add 2 ml of sterile 80% glycerol and mix thoroughly. Transfer 0.5-ml aliquots into 1-ml freezer vials with O-ring seals. Thoroughly shake the freezer vials and freeze at –60°C or lower (typically –80°C). Yeast tend to die if frozen at temperatures above –55°C. Yeast strains can be stored at –80°C indefinitely using this method. Note that strains grown in YPD medium before freezing have better long-term viability than those grown in selective medium.

Retrieval of Frozen Yeast Stocks

Never thaw frozen yeast stocks. Use a sterile loop, sterile wooden stick, or sterile disposable pipette to scratch the surface of the stock. Streak appropriate agar plates (e.g., YPD or selective agar plates) for single colonies. Recap the frozen stock and return it to storage at –80°C. Incubate the plate for 2 days at 30°C. Yeast can be stored for approximately 6 months at 4°C on YPD agar plates or for approximately 2 months on selective plates (i.e., SC plates with added supplements). Plates should be stored upside down at 4°C during this time. For long-term storage, seal the plates or place them in bags to keep them from drying out. Supplementing YPD medium with adenine prevents the toxicity caused by the red pigment produced by *ade2* strains that are stored at 4°C.

Shipment of Yeast Strains by Mail

Patch out clones on an appropriate selective plate. Incubate for 2 days at 30°C. Inoculate 6 ml of YPD medium with a "matchhead-full" of cells from the plate (this is a large inoculum; the culture will be turbid before incubation). Incubate the culture at 30°C with agitation overnight. Add 2 ml of sterile 80% glycerol and mix thoroughly. Transfer 0.5-ml aliquots into 1-ml freezer vials with O-ring seals. Ship the vials by regular first-class mail. On receipt, each cell suspension should be streaked on a suitable agar plate and incubated for 2–3 days at 30°C. Alternatively, ship yeast on YPD plates or in tubes or vials (with loosened caps) containing solid YPD medium (YPD slants).

If *un*saturated cultures are prepared (e.g., by inoculating fresh YPD medium with cells and tightly capping the freezer vial) and sent in the mail, the tubes may explode during transit because of the buildup of pressure from actively fermenting yeast in a tightly sealed vessel.

Yeast strains should only be shipped at room temperature in moderate weather conditions. Hot summer weather may be lethal during shipment.

Mammalian Tissue-culture Cells

Typically, mammalian tissue-culture cells are grown in one of a number of media (e.g., RPMI 1640) supplemented with 5–15% FBS/FCS and 1% L-glutamine. Cultures are grown at 37°C in a 5% CO_2 environment. The volume of medium in the flask can affect the growth of cells, because the surface-to-air ratio is important in maintaining the proper pH of the medium. Factors that can affect the growth characteristics of a cell line include a change in incubation temperature, a difference in the lot of FBS/FCS and/or medium, depletion of L-glutamine in the medium, contamination with mycoplasma, and length of time in continuous culture. To control these factors, medium is warmed to 37°C before it is added to cells, new lots of FBS/FCS and medium are tested for at least 2 weeks with a control cell line before they are accepted for general use, and fresh glutamine is added to the medium as required.

Each cell line has specific growth requirements. The final cell concentration (typically 5×10^6 to 5×10^7 cells/ml) during frozen storage may affect viability. Follow specific handling and storage medium recommendations for each cell line.

DMSO Stocks of Mammalian Cells

Centrifuge freshly grown, healthy cells at 500g at 4°C for 10 minutes. Discard the medium and resuspend the cell pellet in FBS/FCS containing 8–10%

DMSO ▼ at 4°C. Transfer 0.5–1-ml aliquots of cell suspension into cryotubes and place in a –80°C freezer. Samples can be stored at –80°C or transferred into a liquid nitrogen storage tank. Cells should be viable at –80°C for up to 1 year or at –185°C in liquid nitrogen for up to 10 years.

Because each aliquot of frozen mammalian cells should be thawed just before use, at least two separate batches of ten aliquots each should be prepared for each cell line. As a precaution in case of freezer failure, store the aliquots of each cell line in at least two different freezers or liquid nitrogen storage tanks.

Retrieval of Frozen Mammalian Tissue-culture Cells

The DMSO ▼ must first be removed from the frozen cells and the cells revived. To do this, quickly thaw the frozen aliquot of tissue-culture cells by placing it in a water bath set at 37°C with the top of the vessel above the H_2O line. Clean the outside of the vessel with 70% ethanol. Use a sterile pipette to transfer the cells into a 15-ml centrifuge tube containing appropriate medium (typically, 10 ml of medium containing 10% FBS/FCS). Gently centrifuge the cell suspension at 200g at room temperature for 5 minutes. Discard the medium and resuspend the cell pellet in 10 ml of appropriate medium containing FBS/FCS (typically 10%). Transfer the cells into an appropriate-sized flask and incubate at 37°C in a 5% CO_2 environment. Each cell line will recover at a different rate.

Shipment of Mammalian Tissue-culture Cells

Mammalian tissue-culture cell lines can be shipped as frozen stocks or as growing cultures. Frozen stocks should be shipped on dry ice in a Styrofoam container for next-day delivery. To ship growing cultures, inoculate the medium with a small aliquot of cells in a tissue-culture flask so that the cells are approximately a quarter to half confluent on the next day. At the time of shipment, the cells should be in log growth phase but should not be too dense (dense monolayers tend to peel off during transit). Before shipping, fill the flask to the neck with culture medium, cap tightly, and cover the cap with Parafilm M to prevent leaking. Wrap the flask in paper towels or place in a sealable plastic bag and cushion with cotton balls (this prevents the flask from breaking and also absorbs any liquid in case of a leak). Cell lines that grow in suspension can be shipped in centrifuge tubes or flasks filled to the neck and sealed.

Seasonal factors (extreme hot or cold) must be considered when shipping mammalian tissue-culture cells. If extremely warm weather conditions are anticipated, ship live cultures in a Styrofoam container. If extremely cold weather is anticipated, do not ship live cultures. For international shipping, minimize delays at customs by properly stating the value and contents of the package (it is advisable to check with customs officials in advance for additional shipping information). Upon receipt, allow the cells to recover by incubating live cultures overnight at 37°C before unsealing. Additional information about the shipping of tissue-culture cell lines can be obtained from the Coriell Institute for Medical Research, 403 Haddon Avenue, Camden, New Jersey 08103 (http://www.coriell.org; e-mail ccr@coriell.org; phone 1-800-752-3805; fax 856-757-9737).

Blood and Tissue Samples

Blood ▼ to be used for DNA isolation should be collected in EDTA (purple-top Vacutainers). Blood containing EDTA can be stored for 2 months at 4°C. If the white blood cells are to be immortalized, collect the blood in heparin (green-top Vacutainers) to prevent clotting and store at room temperature for a maximum of 4 days. Blood samples can be shipped on wet ice.

Mouse and human tissue samples to be used for DNA preparation should be frozen in liquid nitrogen and stored for up to 6 months at −80°C. Do not thaw samples slowly, because this allows nucleases to degrade the DNA; immediately place them in a denaturing cell lysis solution at the time of DNA isolation. Frozen tissue samples can be shipped on dry ice.

DNA

Impure DNA containing traces of chemicals used during isolation often does not store well. Contamination with heavy metals, free radicals as chemical breakdown products, and oxidation products of phenol degradation can cause breakage of phosphodiester bonds. UV irradiation causes the production of thymine dimers and cross-links, resulting in loss of biological activity. Ethidium bromide causes photooxidation with visible light and molecular oxygen. Nucleases found on human skin do not generally pose a major problem for DNA. (RNases are very stable, but most DNases are not; however, the use of gloves is always recommended.)

Storage of DNA

As a general rule, the more highly purified the DNA, the longer it can be stored under any conditions. DNA in solution is typically stored in TE (pH 8.0). DNA for long-term storage should contain a high salt concentration (at least 1 M

NaCl or other salt) and 10 mM EDTA (to chelate heavy metals). Always dissolve DNA pellets in low-ionic-strength solutions (e.g., TE) and then add more salt if desired. Dried DNA pellets can be stored for up to 6 months at –20°C, and DNA precipitated with ethanol can be stored indefinitely at –20°C.

Storage at 4°C is the best condition for routine storage of highly purified DNA. For storage at –20°C, it is generally preferable to use a non-frost-free freezer. (Single- and double-stranded breaks may occur when the DNA is subjected to frequent freeze/thaw cycles.) For long-term storage, –80°C is recommended.

Shipment of DNA

Highly purified DNA can be shipped as an aqueous solution at room temperature or 4°C (i.e., on ice) or frozen on dry ice. It can also be precipitated with ethanol and then shipped at room temperature either as a dried DNA pellet or as precipitated DNA in ethanol. When the purity of the DNA is in doubt, do not ship DNA at room temperature as an aqueous solution. Trace contamination with nucleases may result in significant degradation of the DNA during shipment.

NOTES

NOTES

SECTION 8
Useful Facts and Tables

Useful Information and Reference Tables

*T*his section provides general reference material from the various Cold Spring Harbor Laboratory Press manuals. A selection of tables of general interest to the bench scientist is presented along with a comprehensive listing of tables found in CSHLP manuals. The subject matter of the tables is organized in the following manner: useful information; DNA, RNA, and proteins (including microarrays and RNAi); strains and vectors, gene expression; imaging; and animal and plant models. This section also includes a set of guidelines for scientific nomenclature and a listing of useful Web sites for scientists.

Printed Tables

Title	Manual code
Abundance classes of mRNA in mammalian and plant cells	MICRO
Components of mammalian cytoplasmic RNA	MICRO
Expected recovery of total RNA from mouse embryos and extraembryonic tissues and fetal and adult tissues	MME3
Absorbance of nucleic acids at 260 nm	MICRO
Concentration of double-stranded DNA in solution	MC3
Amounts of double-stranded DNA required for dideoxy-mediated sequencing reactions	MC3
Primer design: Properties of oligonucleotides that influence the efficiency of amplification	MC3
Commonly used oligonucleotide primers	MC3
Molecular conversions for oligonucleotides	MC3
Theoretical number of cycles required for PCR	MC3
Number of oligonucleotides required to code for all possible peptides of various lengths	MC3
Molar conversions for proteins	MC3
Theoretical R(260/280) values versus percentage protein (%P) and nucleic acids (%N) in mixtures of proteins and nucleic acids	PUR
Concentration limits of interfering compounds for A_{280} and A_{205} protein assays	PUR
Properties of useful sulfhydryl reducing agents	PUR
Inhibitor cocktails used to control proteolysis during protein isolation	PROT
Solvent additives that stabilize or destabilize proteins in solution	PUR
Genomic resources for microarrays	MC3
Absorption properties of fluorescent dyes	MICRO
Names and other identifiers associated with EST clones	MICRO
Mammalian genes targeted by RNAi	RNAi
Human and animal cell lines in which siRNA triggers silencing	RNAi
Size distribution of siRNAs in various eukaryotes	RNAi
Commonly used FPs	LCI
FP-based biosensors	LCI
Physiological indicators using fluorescence resonance energy transfer	IND
Chemical properties of commercially available caged compounds	IND
Summary of the physicochemical properties of caging chromophores	IND
Properties of objective lenses	IND
Objective identifier markings	IND
Optical constants relevant to microscopy	IND
Dimensions of dishes used for cell culture	MC3
Cell-autonomous markers used to follow cell lineages during mouse development	MME3

Abundance Classes of mRNA in Mammalian and Plant Cells MICRO

Total number of mRNA species	$1.0–3.4 \times 10^4$
Sequence complexity	$1–3 \times 10^7$ nucleotides
Total number of mRNA molecules per cell	
High-abundance mRNA species	
Number of species	7–10
Sequence complexity	$\sim1–3 \times 10^4$ nucleotides
Number of molecules/cell	4,500–10,000 molecules/cell
Medium-abundance mRNA species	
Number of species	500–800
Sequence complexity	2×10^5 to 1.5×10^6 nucleotides
Number of molecules/cell	300–1000 molecules/cell
Low-abundance mRNA species	
Number of species	10,000–34,000
Sequence complexity	$1–3 \times 10^7$ nucleotides
Number of molecules/cell	3–100 molecules/cell

It is estimated that ~85% of expressed genes are present at ≤5 copies per cell (<1:100,000); ~97% of the mass of mRNA in a cell contains only ~20% of the transcribed genes.

Components of Mammalian Cytoplasmic RNA MICRO

Type of RNA	Size (nucleotides)	Approximate molecular mass (daltons)[a]	Number of species per cell	Percentage by weight of cytoplasmic RNA
18S ribosomal	1868[b]	0.58×10^6	1	~20
28S ribosomal	5025[b]	1.56×10^6	1	~57
5.8S ribosomal	158	4.9×10^4	1	~1.5
5.0S ribosomal	120	3.70×10^4	1	~1.5
tRNAs	73–91	heterogeneous	~100	15–17
mRNA	several hundred to several thousand	heterogeneous	$>10^4$	1–5
Small RNAs	100–330	heterogeneous	~60	<2

[a]Calculated using a molecular mass of 310 daltons per nucleotide.
[b]The sizes given are those of human 18S and 28S ribosomal RNAs.

Expected Recovery of Total RNA from Mouse Embryos and Extraembryonic Tissues and Fetal and Adult Tissues   MME3

A. Mouse embryo—Stage and tissue	Yield of total RNA
Day 0.5—oocyte	0.4 ng
Day 2.5 (8–16-cell morula)	0.7 ng
Day 3.5 (64-cell blastocyst)	2 ng
Day 6.5—embryo plus ectoplacental cone	0.2 μg
Day 7.5—embryonic region	0.3 μg
Day 7.5—extraembryonic region	0.2 μg
Day 8.5—embryo minus yolk sac	1 μg
Day 8.5—visceral yolk sac	3 μg
Day 9.5—placenta	30 μg
Day 10.5—embryo minus yolk sac	40 μg
Day 11.5—blood	5 μg
Day 12.5—fetus minus yolk sac	260 μg
Day 13.5—fetus	375 μg
Day 13.5—visceral yolk sac	70 μg
Day 13.5—placenta	250 μg
Day 16.5—fetus	2550 μg
Day 16.5—liver	160 μg
Day 17.5—liver	225 μg
Day 17.5—brain	100 μg
Day 17.5—kidney	15 μg

B. Adult tissue (9 weeks of age)	Yield of total RNA
Brain	350 μg
Heart	100 μg
Kidney	425 μg
Spleen	300 μg
Salivary gland	350 μg
Preputial gland (male)	25 μg
Lung	90 μg
Thymus	300 μg
Liver	4500 μg
Ovary	25 μg
Testis	200 μg
Bone marrow (one tibia and one femur)	25 μg
Leg muscle (per hind leg)	200 μg

Information provided by James Lee, Mayo Clinic Scottsdale, S.C. Johnson Medical Research Center, Scottsdale, Arizona 85259.

Absorbance of Nucleic Acids at 260 nm MICRO

Absorbance at 260 nm	Extinction coefficient	Weight of nucleic acid in one OD_{260} unit
RNA and single-stranded DNA	7.4	40 μg/ml
Double-stranded DNA	6.6	50 μg/ml
Oligonucleotides	8.6	33 μg/ml

The values shown in the table are (1) valid for 1-cm path-length cuvettes and (2) should be regarded as approximate because they are affected by the base composition and the degree of secondary structure in single-stranded nucleic acids.

Concentration of Double-stranded DNA in Solution MC3

Double-stranded DNA (50 μg/ml)	BP/molecule	Molecular mass of DNA (daltons)	Molecules DNA/ml	Moles/ml	50 mg/ml Solution Molar concentration of DNA	Phosphate	Termini
Bacteriophage λ	48,514	3.20×10^7	9.41×10^{11}	1.56×10^{-12}	1.56 nM	157 μM	3.12 nM
pAd10SacBII	30,300	2.00×10^7	1.51×10^{12}	2.50×10^{-12}	2.50 nM	157 μM	5.00 nM
pCYPAC1	19,600	1.29×10^7	2.33×10^{12}	3.88×10^{-12}	3.88 nM	157 μM	7.76 nM
pYAC4	11,400	7.52×10^6	4.00×10^{12}	6.65×10^{-12}	6.65 nM	157 μM	13.1 nM
pBeloBACII	7,400	4.88×10^6	6.17×10^{12}	1.03×10^{-11}	10.3 nM	157 μM	20.6 nM
pBR322	4,363	2.88×10^6	1.05×10^{13}	1.74×10^{-11}	17.4 nM	157 μM	34.8 nM
pUC18/pUC19	2,686	1.77×10^6	1.70×10^{13}	2.83×10^{-11}	28.3 nM	157 μM	56.6 nM
Segment of DNA (1 kb)	1,000	6.60×10^5	4.56×10^{13}	7.58×10^{-11}	75.8 nM	157 μM	152 nM
Octameric double-stranded linker	8	5.28×10^3	5.70×10^{15}	9.47×10^{-9}	9.47 μM	157 μM	18.9 nM

Amounts of Double-stranded DNA Required for Dideoxy-mediated Sequencing Reactions MC3

Amount of template required (fmoles)	Size of double-stranded DNA (vector + insert) (kb)				
	1	2.5	5	10	20
10	6.6 ng	16.5 ng	33 ng	66 ng	132 ng
25	16.5 ng	41 ng	82.5 ng	165 ng	330 ng
50	33 ng	82.5 ng	165 ng	330 ng	660 ng
100	66 ng	165 ng	330 ng	660 ng	1.33 µg
150	100 ng	250 ng	475 ng	1.0 µg	2.0 µg
200	133 ng	330 ng	660 ng	1. 33 µg	2.66 µg
250	166 ng	410 ng	850 ng	1.66 µg	3.33 µg
500	330 ng	825 ng	1.65 µg	3.33 µg	6.66 µg

Primer Design: Properties of Oligonucleotides that Influence the Efficiency of *Amplification* MC3

Property	Optimal design
Base composition	G+C content should be between 40% and 60%, with an even distribution of all four bases along the length of the primer (e.g., no polypurine tracts or polypyrimidine tracts and no dinucleotide repeats).
Length	The region of the primer complementary to the template should be 18–25 nucleotides long. Members of a primer pair should not differ in length by >3 bp.
Repeated and self-complementary sequences	No inverted repeat sequences or self-complementary >3 bp in length should be present. Sequences of this type tend to form hairpin structures, which, if stable under PCR conditions, can effectively prevent the sequences oligonucleotide from annealing to its target DNA.
Complementarity between members of a primer pair	The 3′ terminal sequences of one primer should not be able to bind to any site on the other primer. Because primers are present at high concentration in PCR, even weak complementarity between them leads to hybrid formation and the consequent synthesis and amplification of primer dimers. If primer dimers form early in PCR, they can compete for DNA polymerase, primers, and nucleotides and thus can suppress amplification of the target DNA. Formation of primer dimers can be reduced by careful primer design, by the use of hot start or touchdown PCR, and/or by the use of specially formulated DNA polymerases (e.g., AmpliGold, Perkin-Elmer). When more than one primer pair is used in a single PCR, check that none of the 3′ ends have detectable complementarity to any other primers in the reaction.
Melting temperatures (T_m)	Calculated T_m values of members of a primer pair should not differ by >5°C. The T_m of the amplified product should not differ from the T_m values of the primer pairs by >10°C. This property ensures that the amplified product will be efficiently denatured during each cycle of PCR.
3′ Termini	The nature of the 3′ end of primers is crucial. If possible, the 3′ base of each primer should be G or C. However, primers with a ....NNCG or .....NNGC sequence at their 3′ termini are not recommended because the unusually high ΔG of the terminal GC bases promotes the formation of hairpin structures and may generate primer dimers.

(continued)

Primer Design: Properties of Oligonucleotides that Influence the Efficiency of Amplification (continued)

Property	Optimal design
Adding restriction sites, bacteriophage promoters, and other sequences to the 5′ termini of primers	Useful sequences not complementary to the target DNA are commonly added to the 5′ end of the primer. In general, the presence of such sequences does not significantly affect annealing of the oligonucleotide to its target DNA. These additional sequences include bacteriophage promoters and GC clamps. Restriction sites are a special case. Because the efficiency of cleavage of restriction sites located at the 5′ termini of DNA molecules is poor, the primer should be extended by at least three additional nucleotides beyond the recognition sequence of the restriction enzyme. New England BioLabs' catalog contains information on the efficiency with which different restriction enzymes cleave sites near the termini of DNA molecules.
Placement of priming sites	Depending on the purpose of the experiment, the placement of priming sites may be constrained by the location of mutations, restriction sites, coding sequences, microsatellites, or *cis*-acting elements. When designing primers for use on cDNA templates, it is best to use forward and reverse primers that bind to sequences in different exons. This allows amplification products derived from cDNA and contaminating genomic DNA to be easily distinguished.
Primers for degenerate PCR	When a short sequence of amino acids has been obtained by sequencing a purified protein, pools of degenerate oligonucleotides containing all possible coding combinations can be used to amplify the corresponding genomic or cDNA sequences.

Commonly Used Oligonucleotide Primers MC3

Primer	Sequence
λgt10 forward primer	5′-AGCAAGTTCAGCCTGGTTAAG-3′
λgt10 reverse primer	5′-CTTATGAGTATTTCTTCCAGGGTA-3′
λgt11 forward primer	5′-GGTGGCGACGACTCCTGGAGCCCG-3′
λgt11 reverse primer	5′-TTGACACCAGACCAACTGGTAATG-3′
pUC/M13-40 forward primer	5′-GTTTTCCCAGTCACGACG-3′
pUC/M13-48 reverse primer	5′-AGCGGATAACAATTTCACACAGG-3′
pUC/M13-20 forward primer	5′-GTAAAACGACGGCCAGT-3′
pUC/M13-20 reverse primer	5′-GGAAACAGCTATGACCATG-3′
SP6 universal primer	5′-ATTTAGGTGACACTATAG-3′
T7 universal primer	5′-TAATACGACTCACTATAGGG-3′
T3 promoter primer	5′-ATTAACCCTCACTAAAGGGA-3′

Molecular Conversions for Oligonucleotides MC3

Size of oligonucleotide (nucleotides)	Molecular mass (daltons)	Molecules of DNA in 1 μg	Moles of DNA in 1 μg
8	2.64×10^3	2.28×10^{14}	379 pmoles
10	3.30×10^3	1.82×10^{14}	303 pmoles
12	3.96×10^3	1.52×10^{14}	253 pmoles
14	4.62×10^3	1.30×10^{14}	216 pmoles
16	5.28×10^3	1.14×10^{14}	190 pmoles
18	5.94×10^3	1.01×10^{14}	168 pmoles
20	6.60×10^3	9.12×10^{13}	152 pmoles

Theoretical Number of Cycles Required for PCR   MC3

			Targets			
y	1	10	100	1,000	10,000	100,000
1.00	34	30	27	24	20	17
0.95	35	32	28	25	21	18
0.90	36	33	29	26	22	18
0.85	38	34	30	27	23	19
0.60	40	36	32	28	24	20
0.75	42	38	33	29	25	21
0.70	44	40	35	31	27	22
0.65	46	42	37	33	28	23
0.60	49	45	40	35	30	25
0.55	53	48	43	37	32	27
0.50	57	52	46	40	35	29
0.45	62	56	50	44	38	31
0.40	69	62	55	48	42	35
0.35	77	70	62	54	47	39
0.30	88	79	71	62	53	44
0.25	104	93	83	73	62	52
0.20	127	114	102	89	76	64
0.15	165	149	132	116	99	83
0.10	242	218	194	170	145	121

Reprinted, with permission of Springer Science and Business Media, from J. Rameckers et al. 1997. How many cycles does a PCR need? Determinations of cycle numbers depending on the number of targets and the reaction efficiency factor. *Naturwissenschaften 84:* 259–262 (©Springer-Verlag).

Number of PCR cycles (rounded) required to reach 10 ng of DNA (based on a 200-bp PCR product) at various efficiency levels (y) and various target numbers (targets).

Number of Oligonucleotides Required to Code for All Possible Peptides of Various Lengths   MC3

Length of peptide (amino acid residues)	Number of oligonucleotides in degenerate pool
4	$20^4 = 1.6 \times 10^5$
5	$20^5 = 3.2 \times 10^6$
6	$20^6 = 6.4 \times 10^7$
7	$20^7 = 1.28 \times 10^9$
8	$20^8 = 2.56 \times 10^{10}$
10	$20^{10} = 1.024 \times 10^{13}$
12	$20^{12} = 2.028 \times 10^{14}$

Molar Conversions for Proteins MC3

Molecular weight of unmodified protein	Approximate number of residues	Molecules of protein in 1 ml of a solution containing 1 mg/ml	Moles of protein in 1 ml of a solution containing 1 mg/ml	Molar concentration of protein solution containing 1 mg/ml
100,000	917	6×10^{15}	10^{-8}	10^{-5} M
80,000	734	7.5×10^{15}	1.25×10^{-8}	1.25×10^{-5} M
60,000	550	10^{16}	1.66×10^{-8}	1.66×10^{-5} M
40,000	367	1.5×10^{16}	2.50×10^{-8}	2.5×10^{-5} M
20,000	183	3×10^{16}	50×10^{-8}	5.0×10^{-5} M
10,000	92	6×10^{16}	10^{-7}	10^{-4} M

Theoretical R(260/280) Values Versus Percentage Protein (%P) and Nucleic Acids (%N) in Mixtures of Proteins and Nucleic Acids PUR

%P	%N	R(260/280)
100	0	0.57
95	5	1.06
90	10	1.32
85	15	1.48
80	20	1.59
75	25	1.67
70	30	1.73
65	35	1.78
60	40	1.91
55	45	1.84
50	50	1.87
45	55	1.89
40	60	1.91
35	65	1.93
30	70	1.95
25	75	1.95
20	80	1.97
15	85	1.98
10	90	1.98
5	95	1.99
0	100	2.00

Reprinted, with permission, from J.A. Glasel 1995. Validity of nucleic acid purities monitored by 260nm/280nm absorbance ratios. *BioTechniques 18:* 62–63.

Concentration Limits of Interfering Compounds for A_{280} and A_{205} Protein Assays PUR

Compound[a]	Concentration limits (nm)	
	280	205
Acids and bases		
HCl	>1 M	0.5 M
NaOH	>1 M	2.5 mM
PCA	>10%	1 M
TCA	>10%	<1%
Buffers		
acetate	0.1 M	10 mM
ammonium sulfate	>50%	9%
borate		>100 mM
citrate	5%	<10 mM
glycine	1 M	5 mM
HEPES	–	<20 mM
phosphate	1 M	50 mM
Tris	0.5 M	40 mM
Detergents		
Brij 35	1%	1%
CHAPS	10%	<0.1%
deoxycholate	0.30%	0.1%
digitonin	10%	–
lubrol PX	10%	–
octylglucoside	10%	–
SDS	0.10%	0.10%
Triton X-100	0.02%	0.01%
Triton X-100 (reduced)	>10%	2%
Tween-20	0.30%	0.10%
Reductants		
dithiothreitol	3 mM	0.1 mM
2-mercaptoethanol	0 mM	<10 mM
Miscellaneous		
DNA/RNA	1 μg	–
DMSO	20%	<10%
EDTA	30 mM	0.2 mM
glycerol	40%	5%
KCl	100 mM	50 mM
NaCl	>1 M	0.6 M
sucrose	10 M	0.5 M
urea	>1 M	<0.1 M
azide	0.01%	–

[a](PCA) Perchloric acid; (TCA) trichloroacetic acid; (HEPES) N-2-hydroxy-ethylpiperazine-N'-2-ethanesulfonic acid; (CHAPS) (3-[(cholamidopropyl) dimethyl-ammonio]-1-propane-sulfonate); (SDS) sodium dodecyl sulfate; (DMSO) dimethylsulfoxide; (EDTA) ethylenediamine tetraacetic acid.

Properties of Useful Sulfhydryl Reducing Agents PUR

Sulfhydryl reagent	Molecular weight	Density of liquid	Molarity of liquid	Comments
2-Mercaptoethanol[a]	78	1.117 g/ml	14.3 M	volatile
Thioglycerol	108	1.295 g/ml	12 M	
Dithiothreitol or dithioerythreitol	154	solid		strong reducing agents
2,3-Dimercaptopropanol	124	1.239 g/ml	10 M	moderately soluble in water
Tributylphosphine[a]	203	0.81 g/ml		slightly soluble in water
tris(2-carboxyethyl) phosphine (TCEP)				soluble in water; strong reductant over a wide pH range (1.5–8.5); stronger reductant than DTT at pH values <8.0.
L-Cysteine	121	solid		
Thioglycolic acid	92	solid		
Glutathione (reduced)	307	solid		

[a]Because of high toxicity and offensive odors, use in a chemical fume hood.

Inhibitor Cocktails Used to Control Proteolysis During Protein Isolation PROT

Tissue type	Protease inhibitors (working concentration)	Target protease type	Stock solution[a]
Animal tissues	AEBSF (0.2 mM) (or DCI [0.1 mM] or PMSF [0.2 mM])	serine	20 mM in methanol (DCI: 10 mM in DMSO; PMSF: 200 mM in ethanol or isopropanol)
	benzamidine (1 mM)	serine	100 mM
	leupeptin (10 µg/ml)	serine/cysteine	1 mg/ml
	pepstatin (10 µg/ml)	aspartic	5 mg/ml in methanol
	aprotinin (trasylol)(1 µg/ml)	serine	0.1 mg/ml
	EDTA or EGTA[b] (1 mM)	metallo	100 mM
Plant tissues	AEBSF (0.2 mM) (or DCI [0.1 mM] or PMSF [0.2 mM])	serine	20 mM in methanol (DCI: 10 mM in DMSO; PMSF: 200 mM in ethanol or isopropanol)
	chymostatin (20 µg/ml)	serine/cysteine	1 mg/ml in DMSO
	EDTA or EGTA[b] (1 mM)	metallo	100 mM
Yeast and fungi	AEBSF (0.2 mM) (or DCI [0.1 mM] or PMSF [0.2 mM])	serine	20 mM in methanol (DCI: 10 mM in DMSO; PMSF: 200 mM in ethanol or isopropanol)
	pepstatin (15 µg/ml)	aspartic	5 mg/ml in methanol
	1,10-phenanthroline (5 mM)	metallo	1 M in ethanol
Bacteria	AEBSF (0.2 mM) (or DCI [0.1 mM] or PMSF [0.2 mM])	serine	20 mM in methanol (DCI: 10 mM in DMSO; PMSF: 200 mM in ethanol or isopropanol)
	EDTA or EGTA[b] (1 mM)	metallo	100 mM

Abbreviations: (AEBSF) 4-(2-aminoethyl)-benzenesulfonylfluoride; (DCI) 3,4-dichloroisocoumarin; (DMSO) dimethylsulfoxide; (EDTA) ethylenediamine tetraacetic acid; (EGTA) ethylene glycol bis (β-aminoethyl ether) N,N,N´,N´-tetraacetic acid; (PMSF) phenylmethylsulfonyl fluoride. M_r values of inhibitors: AEBSF (240); PMSF (174); DCI (215); EDTA (disodium salt, dihydrate) (372); benzamidine (hydrochloride) (157); leupeptin (427); pepstatin (686); aprotinin (6500); chymostatin (605); 1,10-phenanthroline (198).
[a]Aqueous solution unless otherwise indicated.
[b]An efficient chelator of divalent metal cations other than Mg^{++} (for which it has a 10^{-3}-fold lower affinity).

Solvent Additives that Stabilize or Destabilize Proteins in Solution PUR

Compounds	Mode of action	Working concentration
Osmolytic stabilizers[a]		
Polyols and sugars		
glycerol, erythritol, arabitol, sorbitol, mannitol, xylitol, mannisidomannitol, glucosylglycerol, glucose, fructose, sucrose, trehalose, isofluoroside	These stabilize the lattice structure of the water, thereby increasing surface tension and viscosity. In addition, they stabilize hydration shells and protect against aggregation by increasing the molecular density of the solution without changing the dielectric constant.	10–40%
Polymers		
dextrans, levans, polyethylene glycol	Polymers increase molecular density and solvent viscosity, thus lowering protein aggregation in a single-phase system. At high polymer concentration, a two-phase system develops and the protein aggregates in the phase in which its concentration is highest.	1–15%
Amino acids and their derivatives		
glycine, alanine, proline, taurine, betaine, octopine, glutamate, sarcosine, γ-amino-butyric acid, trimethylamine N-oxide (TMAO)	Small amino acids with no net charge, such as glycine and alanine, have weak electrostatic interactions with proteins. Octopine is a derivative of arginine that is less denaturing to proteins. TMAO stabilizes proteins even in the presence of denaturants such as urea. Most of these compounds increase the surface tension of water.	20–500 mM

Ionic stabilizers[b]

Salts

citrate, sulfates, acetate, phosphates, quaternary amines	Larger anions shield charges and can stabilize proteins at low concentrations. At high concentrations, they lead to precipitation due to competition for water molecules.	20–400 mM

Ionic destabilizers

Salts

chlorides, nitrates, thiocyanates	Although generally less stabilizing than large ions (especially polyvalent ions), these are useful for charge shielding at lower concentrations.	20–400 mM

Denaturants (chaotrophs)

urea, guanidinium salts, trichloroacetates, cetylmethyl-ammonium salts, organic solvents	Denaturants either stabilize the unfolded state of proteins (urea) or perturb protein structure by interfering with hydrogen bonding or disturbing the hydration shell.	0.2–8 M

Adapted, in part, with permission from Macmillan Publishers Ltd., from C.H. Schein. 1990. Solubility as a function of protein instruction and solvent components. *Bio/Technology* 8: 308–317.

[a]In general, the osmolytic stabilizers have little direct interaction with proteins, but affect the bulk solution properties in water.

[b]These affect enzyme reactions. Their stabilizing effects on proteins occur within a concentration range much narrower than that of osmolyte stabilizers.

Genomic Resources for Microarrays MC3

Organism	Material	Comments	Resource
Rat	6000 cDNA clones	1500 sequence-validated clones are available.	Invitrogen (www.invitrogen.com)
Yeast (S. cerevisiae)	ORF-specific primer pairs	These primers are designed to amplify complete coding regions including start and stop codons from genomic DNA (possible because very few yeast genes contain introns). Recently, The Sanger Centre and the Imperial Cancer Research Fund have initiated work on amplification of coding regions from the fission yeast S. pombe.	Invitrogen (www.invitrogen.com)
	intergenic region primer pairs	These primer pairs are designed to amplify regions that lie between the open reading frames (ORFs), from genomic DNA, for use in cloning yeast promoters upstream of reporter genes, assaying deletions or insertions in yeast genes, and mapping transcription-factor-binding sites.	Invitrogen (www.invitrogen.com)
Arabidopsis thaliana	7900-member clone set	The clones represent genes from all major tissue categories (e.g., roots, rosettes, and inflorescence).	Incyte Genomics (www.incyte.com/products/organisms.html)
	11,500 EST clones	Collection was generated at Michigan State University.	Affymetrix (www.affymetrix.com)
Caenorhabditis elegans	cDNA arrays representing the complete set of C. elegans genes	These microarrays along with support for hybridization and data analysis are available to other C. elegans laboratories on a collaborative basis.	Stuart Kim's laboratory at Stanford University (cmgm.stanford.edu/~kimlab/)
	primer pairs	These primer pairs can amplify all or a portion of each of the 19,000 genes.	Invitrogen (www.invitrogen.com)

Drosophila melanogaster	cDNAs representing 12,000 genes	The cDNA collection is currently undergoing annotation at the Berkley *Drosophila* Genome Project.	Berkeley *Drosophila* Genome Project (www.fruitfly.org) will be available for purchase through Invitrogen (www.invitrogen.com)
Escherichia coli	4290 ORF sequences	Membrane arrays containing a complete set of PCR-amplified genes.	Genosys Biotechnologies (http://www.sigma-genosys.com/index.asp
Bacillus subtilis	4107 ORF sequences	PCR-amplified ORFs.	Genosys Biotechnologies (www.sigma-geosys.com/index.asp)
Helicobactor pylori	putative 1590 ORFs of strain 26,695; 91 ORFs unique to strain J99		Genosys Biotechnologies (www.sigma-geosys.com/index.asp)
C. jejuni	1654 genes		Genosys Biotechnologies (www.sigma-geosys.com/index.asp)
In Development			
Staphylococcus aureus	1900 ORF sequences		Incyte Genomics (www.incyte.com/products/organisms.html)
Candida albicans	ORF sequences		Incyte Genomics (www.incyte.com/products/organisms.html)

Absorption Properties of Fluorescent Dyes   MICRO

Dye	λ_{max} (nm)		ε (cm^{-1}m^{-1})	Correction factor (CF$_{260}$)
Alexa Fluor 488	492	(green)	62,000	0.30
Alexa Fluor 532	525	(yellow)	82,300	0.24
Alexa Fluor 546	555	(orange)	104,000	0.21
Alexa Fluor 568	576	(red)	93,000	0.45
Alexa Fluor 594	588	(red)	80,400	0.45
Cy3	550	(orange)	150,000	0.08
Cy5	649	(far red)	250,000	0.05

Names and Other Identifiers Associated with EST Clones MICRO

Identifier	Format	Correspondence to clone	Example
ATCC number	5–7 digits	unique	105303
MGC number	4–7 digits	unique	20317
Clone name or alias	alphanumeric	may be more than one per clone	NIB345
IMAGE Clone ID	5–7 digits	may be more than one per clone	60800
GenBank accession number	capital letter followed by 5 digits or 2 letters followed by 6 digits	may be more than one per clone	T12121, AA234567
EST number	EST followed by digits	0 to several per clone	EST149990
Washington University sequence ID	lowercase alphanumeric string followed by a period, then r or s, then a digit	0–2 per clone	Yb34g01.s1
Unigene	2 letters followed by 1–6 digits	unique	Hs.1254

Mammalian Genes Targeted by RNAi RNAi

Gene	Function	Method	Tissue	Silencing[a] mRNA	Silencing[a] Protein	Phenotype
E-cadherin	adhesion	dsRNA	embryo	n.d.	reduced	yes
Egfp (transgene)	fluorescent marker	dsRNA	embryo	n.d.	reduced	yes
Mos	kinase	dsRNA, plasmid	oocyte	reduced	reduced	yes
Plat	protease	dsRNA	oocyte	reduced	reduced	yes
Msy2	RNA binding	dsRNA	oocyte	reduced	n.d.	n.d.
CaMKII	Ca dep. kinase	dsRNA	oocyte	reduced	reduced	n.d.
Iptr1	Ca channel	dsRNA	oocyte	reduced	reduced	yes
Basonuclin	transcription factor	dsRNA	oocyte	reduced	n.d.	n.d.
Acrogranin	cell adhesion	dsRNA	embryo	reduced	n.d.	no
MDicer	dsRNase	dsRNA, siRNA	embryo	reduced	reduced	yes
NuMA	nuclear protein	siRNA	HeLa	n.d.	reduced	yes
GAS41	nuclear protein	siRNA	HeLa	n.d.	reduced	yes
SV40 T antigen	nuclear protein	siRNA	rat fibroblast	n.d.	reduced	n.d.
Lamin A/C	nuclear envelope	siRNA	HeLa	n.d.	reduced	yes
Lamin B1	nuclear envelope	siRNA	HeLa	n.d.	reduced	yes
Lamin B2	nuclear envelope	siRNA	HeLa	n.d.	reduced	yes

LAP2	nuclear envelope	siRNA	HeLa	n.d.	reduced	n.d.
Emerin	nuclear envelope	siRNA	HeLa	n.d.	reduced	n.d.
Nup153	nuclear envelope	siRNA	HeLa	n.d.	reduced	n.d.
β-actin	cytopl. cytoskeleton	siRNA	HeLa	n.d.	reduced	yes
γ-actin	cytopl. cytoskeleton	siRNA	HeLa	n.d.	n.d.	yes
ARC21	cytopl. cytoskeleton	siRNA	HeLa	n.d.	reduced	n.d.
VASP	cytopl. cytoskeleton	siRNA	HeLa	n.d.	reduced	n.d.
Vinculin	cytopl. cytoskeleton	siRNA	mouse 3T3	n.d.	reduced	n.d.
Zyxin	cytopl. cytoskeleton	siRNA	mouse 3T3	n.d.	reduced	yes
Vimentin	cytopl. cytoskeleton	siRNA	HeLa	n.d.	reduced	n.d.
Keratin18	cytopl. cytoskeleton	siRNA	HeLa	n.d.	reduced	n.d.
Eg5	cytopl. cytoskeleton	siRNA	HeLa	n.d.	n.d.	yes
CENP-E	centromere	siRNA	HeLa	n.d.	n.d.	yes
cytopl.dynein	spindle	siRNA	HeLa	n.d.	n.d.	yes
CdK1	kinase	siRNA	HeLa	n.d.	n.d.	yes
Mad2	spindle checkpoint	siRNA	HeLa	n.d.	reduced	yes
Mst1	signaling	siRNA	HeLa	n.d.	no	no

[a]n.d. indicates not determined.

Human and Animal Cell Lines in which siRNA Triggers Silencing   RNAi

Cell line	Tissue origin
A-431	human epidermoid carcinoma
A549	human lung carcinoma
BV173	human B-precursor leukemia
C-33A	human papillomavirus-negative cervical carcinoma
CA46	human Burkitt's lymphoma
Caco2	human colon epithelial cells
CHO	Chinese hamster ovary
COS-7	African green monkey kidney
F5	rat fibroblast
H1299	human nonsmall cell lung carcinoma
HaCaT	human keratinocyte cell
HEK 293	human embryonic kidney
HeLa	human papillomavirus-positive cervical carcinoma
Hep3B	human hepatocellular carcinoma
HUVEC	human umbilical vein endothelial cells
IMR-90	human diploid fibroblast
K562	human chronic myelogenous leukemia, blast crisis
Karpas 299	human T-cell lymphoma
MCF-7	human breast cancer
MDA-MB-468	human breast cancer
MV-411	human acute monocytic leukemia
NIH-3T3	mouse fibroblast
P19	mouse embryonic carcinoma
SD1	human acute lymphoblastic leukemia
SKBR3	human breast cancer
U2OS	human osteogenic sarcoma cell

Size Distribution of siRNAs in Various Eukaryotes RNAi

Organism	Predominant length of siRNA (nucleotides)
Plants	21–23
Neurospora crassa	25[a]
Drosophila melanogaster	21–22
Caenorhabditis elegans	23
Trypanosoma brucei	24–26
Mus musculus	21–22

[a]It is likely that the siRNA length was overestimated because DNA size markers that migrate faster than RNA size markers were used for analysis.

Commonly Used FPs LCI

Species	FP	Excitation (nm)	Emission (nm)	Mutations
Aequorea victoria				
Green	wtGFP	395	475	
	EGFP	489	508	F64L, S65T, H231L
	GFP	399	511	F64L
Blue	EBFP	383	445	F64L, S65T, Y66H, T145F
Cyan	ECFP	434 (453)	477 (501)	F64L, S65T, Y66W, N146I, M153T, V163A, H231L
	CGFP	458	504	ECFP mutations, T203Y
Yellow	EYFP	514	527	S65G, V68L, S72A, T203Y, H231L
	Citrine	516	529	EYFP + Q69M
	Venus	515	528	EYFP + F46L, F64L, M153T, V163A, S175G
Monomeric				A206K, L221K, F223R
Discosoma striata				
Red	DsRed	558	583	
	DsRed2	558	583	R2A, K5E, K9T, A105V, I161T, S197A
	DsRed-Express	554	586	R2A, K5E, N6D, T21S, H41T, N42Q, V44A, C117S, T217A
	RedStar	558	583	2S (insertion), R17K, V96I, F124L, M182K, P186Q, T202I
Monomeric	mRed1	584	607	>30 amino acid substitutions

FP-based Biosensors LCI

Category	Variant	Comments
FRET-based sensors		
Ca^{++}	cameleon	$K_d1 = 100$ nM, $K_d2 = 4.3$ μM, 1.6-fold dynamic
	cameleon (YC6.1)	$K_d = 110$ nM, twofold dynamic range
Phosphorylation	protein kinase A	
	Src	
	Abl	
	EGF receptor	
	insulin receptors	
Miscellaneous	nitric oxide	
	voltage (VSFP)	
	cAMP	
	cGMP	
Sensitized FPs		
Ca^{++}	camgaroo-2	Brightens with Ca^{++}, $K_d = 5$ μM
	G-CaMP	Brightens with Ca^{++}, $K_d = 235$ nM, folds at 28°C
	pericams:	Folds at 37°C, $K_d = 1.2$ μM
	flash	Brightens with Ca^{++}
	inverse	Dims on binding Ca^{++}
	ratiometric	Changes excitation from 420 to 490 nm with Ca^{++}, tenfold dynamic range

(continued)

FP-based Biosensors *(continued)*

Category	Variant	Comments
pH	tandem EGFP:ECFP:EYFP	Uses pH dependency of FPs to detect pH 5–8
	pHluorins:	pKa = 7.1
	ratiometric	Changes excitation from 470 to 410 nm in acid, 2.5-fold dynamic range
	ecliptic	Fluorescence quenches in acid
	deGFP4	pKa = 7.4, changes emission from 518 nm to 461 nm (400-nm excitation) in acid, fivefold dynamic range
Cl⁻	EYFP	
	tandem ECFP:EYFP	
Voltage	FlaSH	
Molecular highlighters		
	PA-GFP	100-fold increase in fluorescence after 413-nm illumination
	DsRed	Converts to green emission by three-photon excitation (<760 nm)
	pTimer	Converts from green (483-nm excitation; 500-nm emission) to red (558-nm excitation; 583-nm emission) over time
	Kaede	Emission shifts from 518 to 582 nm (480 excitation) when exposed to ultraviolet light
	KFP1	Reversible or irreversible fluorescence (580-nm excitation; 600-nm emission) using green light; Blue light quenches reversible fluorescence

Physiological Indicators Using Fluorescence Resonance Energy Transfer IND

Analyte or process	Donor and acceptor[b]	ΔFRET[c]	Wavelength (nm)[a]			Maximum emission ratio changed[d]
			Donor excitation	Donor emission	Acceptor emission	
Depolarization	coumarin-phosphatidylethanolamine;	→	414	450	560	1.8/(100 mV)
	bis(thiobarbiturate)trimethineoxonol	→	495	520	580	1.6–2.2
cAMP	PKA catalytic subunit-FITC; PKA regulatory subunit-TROSu[e]					
Trypsin	BFP-(trypsin-sensitive linker)-GFP	→	380	445	507	4.6
Factor X_a	BFP-(factor X_a-sensitive linker)-GFP	→	385	450	505	1.9
Caspase-3	BFP-(caspase-3-sensitive linker)-GFP	→	380	440	511	?
Ca^{++}-CaM	GFP-CB$_{SM}$-BFP	→	380	448	505	5.7
Ca^{++}	GFP-CB$_{SM}$-BFP-CaMCN	→	380	440	505	1.67
Ca^{++}	ECFP-CaM-M13-EYFP	→	433	476	527	2.1
Ca^{++}	ECFP-CaM; M13-EYFP	←	433	476	527	4
β-Lactamase expression	coumarin-cephalosporin-fluorescein (CCF2)	→	409	447	520	70

[a]Donor excitation wavelength, donor emission wavelength, and acceptor emission wavelength are all in nanometers. Small differences (up to 8 nm) in the wavelengths cited by different laboratories for BFP and GFPs are probably not significant.

[b]Hyphens indicate covalent conjugation or fusion. Interacting donor and acceptor molecules are separated by semicolons.

[c]FRET indicates whether the efficiency of fluorescence resonance energy transfer is increased (↑) or decreased (↓) by the analyte or process.

[d]Maximum factor by which emission ratio changes from zero to saturating levels of the analyte or process, except for depolarization of 100-mV amplitude.

[e]Tetramethylrhodamine N-hydroxysuccinimide.

Chemical Properties of Commercially Available Caged Compounds IND

Cage	Quantum yield	Uncaging rate (sec^{-1})	Source
DM-nitrophen	0.18	3.8×10^4	CB, MP
NP-EGTA	0.23	6.8×10^4	MP
nitr-5	0.04	n.d.	CB, Sigma-Aldrich
CNB-glu	0.14	4.8×10^4	MP
CNB-GABA	0.16	3.6×10^4	MP
CNB-NMDA	0.43	32 and 1.7×10^3	MP
MNI-glu	0.085	$\sim 10^6$	Tocris
NPE-IP3	0.65	225 and 280	MP, CB
DMNB-fluorescein	n.d.	n.d.	MP
NPE-cADPribose	0.11	18	MP
NPE-ATP	0.48 or 0.63	90	CB, MP
diazo-2	0.03	2.3×10^3	MP
CNB-carbamoylcholine	0.8	1.7×10^4	MP

Abbreviations: CB = Calbiochem; MP = Molecular Probes; n.d. = no data.

Summary of the Physicochemical Properties of Caging Chromophores IND

Cage	Absorption max (nm) (extinction/$mM^{-1}cm^{-1}$)	Fluorescence	QY (chem)	QY (fluo)	2PCS (GM)	Rate (sec^{-1})
NB[a,b]	(~0.9 at 350 nm)	none	0.04–0.7	none	0.001	100–10^6
DMNB[b]	350 (5.2)	none	0.01–0.23	none	0.01	140–80,000
7-MCM	328 (13.3)	400	0.1	0.03	n.d.	10^8
DEACM	399 (19.5)	484	0.22	0.0058	n.d.	10^8
DMCM	349 (11.3)	444	0.04	0.022	n.d.	10^8
pHP[a,c]	very pH dependent	n.d.	0.1–0.7	n.d.	n.d.	10^6–10^8
Bhc[c]	370 (15.0)	530	0.012–0.04	0.2–0.4	0.95	n.d.[d]
BHQ[c]	369 (2.6)	510	0.29	0.04	0.54	n.d.[d]
MNI	335 (3.6)	none	0.085	none	0.06	>10^5

Abbreviations and notes: QY = quantum yield (chemical or fluorescent); 2PCS = two-photon cross section at 720–740 nm; n.d. = no data; NB = nitrobenzyl; DMNB = dimethoxynitrobenzyl; 7-MCM = 7-methoxycoumarinylmethyl; DEACM = 7-diethylaminocoumarinylmethyl; DMCM = 6,7-dimethoxycoumarinylmethyl; pHP = *para*-hydroxyphenacetyl; Bhc = 6-bromo-7-hydroxycoumarinylmethyl; BHQ = 8-bromo-7-hydroxyquinolino; MNI = 4-methoxy-7-nitroindolino. Values for the properties of 7-MCM, DEACM, and DMCM are for caged cAMP, Bhc, and MNI for caged glutamate, BHQ for "caged acetate."

[a] Absorption maxima are in the UV range (~280 nm) for both NB and pHP.

[b] The NB and DMNB are of such wide applicability as caging chromophores that it is impossible to make general statements as to their QY and rate of uncaging because these properties vary greatly depending on the type of functionality and compound caged. In contrast, the other cages summarized in the table have been developed for phosphate and carboxylic acids and are *only* useful for these functionalities.

[c] The absorption maxima of these cages are very pH sensitive because they are phenols that deprotonate in the physiological range; e.g., pHP has a pK_a of 8.

[d] Probably uncage with a high rate, by analogy with other coumarin cages.

Properties of Objective Lenses IND

Objective lens	Magnification	NA	WD (mm)	Comments	Manufacturer[a]
Achromat	4x	0.1	16		N
	10x	0.25	6.1		O
	20x	0.40	3.0		O
	40x	0.65	0.45		O
	60x	0.80	0.15		O
	100x	1.25	0.13	oil immersion	O
Plan Achromat	0.5x	0.02	7.0		N
	1x	0.04	3.2		N
	2x	0.05–0.06	5–7.5		N,O
	4x	0.10	22–30		N,O
	10x	0.25	10.5		N,O
	20x	0.40	1.30		N,O
	40x	0.65	0.56		N,O
	50x	0.90–0.50	0.20–0.40	oil immersion	N,O
	100x	1.25	0.15–0.17	oil immersion	N,O
Plan Fluor (Neofluor)	1.25x	0.04	3.5		Z
	2.5x	0.075	9.3		Z
	4x	0.16	13.0		N,O
	5x	0.15	13.6		Z
	10x	0.30	5.6–10.0		N,O,Z
	20x	0.50	1.3–1.6		N,O,Z
	40x	0.75	0.33–0.51		N,O,Z
	60x	1.25	0.10		N,O

		NA	WD		
	63x	1.25	0.10	oil immersion	Z
	100x	1.30	0.05–0.10	oil immersion	N,O,Z
Epiplan-Neofluor	1.25x	0.035	3.0		Z
	5x	0.15	13.7	darkfield WD = 5.5	Z
	10x	0.30	5.7	darkfield WD = 3.5	Z
	20x	0.50	1.4	darkfield WD = 1.2	Z
	20x	0.65	0.29	oil, polarizing	Z
	50x	0.80	0.58		Z
	50x	1.0	0.29	oil, polarizing	Z
	100x	0.90	0.24	darkfield WD = 0.27	Z
	100x	1.3	0.13	oil, polarizing	Z
Plan-Apochromat	1.25x	0.04	5.10		O
	2x	0.08	6.20		N,O
	4x	0.16	12.2		N,Z
	10x	0.32	1.9		N,Z
	10x	0.45	2.8	optimum resolution	N,Z
	20x	0.60	0.45		N,Z
	20x	0.75	0.61	optimum resolution	N,Z
	40x	0.95	0.13–0.16	correction	N,O,Z
	40x	1.00	0.31	oil immersion	N,Z
	60x	1.40	0.10	oil immersion	N,O
	63x	1.40	0.09	oil immersion	Z
	100x	1.40	0.09–0.10	oil immersion	N,O,Z

[a]This information was collected from Zeiss (Z), Olympus (O), and Nikon (N). NA indicates the numerical aperture of the lens; WD indicates the working distance of the lens. For specialty objectives, contact the respective manufacturers.

Objective Identifier Markings   IND

Identifier	Symbol	Typical value/explanation
Degree of correction		
Achromat	—	basic correction for chromatic aberration (CA)
Fluar	—	moderate correction for CA
Apochromat	—	high correction for CA
Plan	—	flat-field correction
Magnification	(M, x)	0.5–40x (dry), 10–100x (water), 40–100x (oil)
Numerical aperture	(NA)	0.02–0.75 (dry), 0.3–1.2 (water), 0.5–1.4 (oil)
Immersion type		
dry	(no marking)	
water (direct)	W, WI	
water (cover glass)	W Korr	adjustable correction for cover-glass thickness
glycerol	G, Glyc	fused silica cover glass—0.2 mm
oil	Oil, Oel	
multi-	Imm	adjustable correction for water, glycerol, oil
Tube length/cover-glass thickness		
160-mm body tube	160/0.17	standard cover glass
	160/—	unspecified or none
infinite conjugate (IC)	∞/0.17	standard cover glass
	∞/—	unspecified or none
Specialized use		
phase contrast	Ph1, Ph2, Ph3	standard phase annuli/phase plates
polarized light	Pol, DIC	strain-free lens elements
UV fluorescence	U-, U340/380	UV transmissive lens elements
dark field	iris	internal iris for variable NA

Optical Constants Relevant to Microscopy IND

Visible light wavelength band (λ)		390–750 nm
Speed of light in free space (c)		3.0×10^8 m/sec (30 cm/nsec)
Frequency band of visible light		$6.7 – 4.3 \times 10^{14}$ Hz
Important wavelengths		
Peak human scotopic vision	556 nm	(yellow green)
Mercury green lamp	546.1 nm	standard optical design wavelength
Low-pressure sodium lamp	589 nm	standard yellow wavelength for refractometry
(Doublet)	589.6 nm	
Red helium-neon laser	632.8 nm	
Important refractive indices (n)		
Air	1.00028	
Water	1.333	(20°C)
Culture medium or saline	1.335	
Animal cells	1.36	(average, by refractometry)
Glycerol	1.47	
Silica	1.46	
Crown glass	1.52	

Average refractive index increment for proteins and nucleic acids: 0.0018 per gm/deciliter

Dimensions of Dishes Used for Cell Culture   MC3

Size of plate	Growth area (cm²)	Relative area[a]	Recommended volume
96 well	0.32	0.04x	200 μl
24 well	1.88	0.25x	500 μl
12 well	3.83	0.5x	1.0 ml
6 well	9.4	1.2x	2.0 ml
35 mm	8.0	1.0x	2.0 ml
60 mm	21	2.6x	5.0 ml
10 cm	55	7x	10.0 ml
Flasks	25	3x	5.0 ml
	75	9x	12.0 ml

[a]Relative area is expressed as a factor of the growth area of a 35-mm culture plate.

Cell-autonomous Markers Used to Follow Cell Lineages during Mouse Development   MME3

A. *Dyes, enzymes, or nucleic acids for microinjection*
 1. DiI, DiO
 2. Lysinated dextran rhodamine and rhodamine-conjugated dextran
 3. Horseradish peroxidase
 4. *EGFP* mRNA

B. *Transgenic mouse lines*
 1. Line carrying about 1000 copies of the β-globin gene; ES cell lines are available
 2. *lacZ* driven by a constitutive promoter
 3. Gene trap lines (e.g., *Gtrosa26*)
 4. Human placental alkaline phosphatase
 5. Fluorescent proteins (FP)
 6. Homologous recombination of inactive duplication in *lacZ* driven by *hprt*
 7. Conditionally expressed *lacZ*, AP, and FP reporters
 8. Replication-defective retrovirus containing *lacZ* reporter

C. *Genetic differences between inbred mouse strains*
 1. Monoclonal antibodies specific for H-2^b and H-2^k
 2. Satellite DNA sequence distribution between) *M. musculus* and *M. caroli*
 3. Null mutation in cytoplasmic malic enzyme (*Mod-1*^{a} vs. *Mod-1*^{b})
 4. Carbohydrate polymorphism recognized by *Dolichos biflorus* agglutinin
 5. Y-specific probe in XX-XY chimeras
 6. Monoclonal antibody (OX7) specific for the Thy-1.1 allele of Thy-1, a surface glycoprotein on T lymphocytes and certain other tissues, including fibroblasts and embryonic brain cells; has been used to follow grafts of embryonic tissue into adult brain; congenic pairs of Thy-1.1 and Thy-1.2 mice are available
 7. High- and low-activity alleles of β-glucuronidase; use limited to central nervous system

Comprehensive Table Key

[a]Refer to page viii for key to manual codes.

Section 2: DNA, RNA, and Proteins

I. NUCLEIC ACIDS

A. Vital Statistics

B. Electrophoretic Separation of Nucleic Acids

Comprehensive Table Key *(continued)*

Comprehensive Table Key *(continued)*

Comprehensive Table Key (*continued*)

Comprehensive Table Key (*continued*)

Comprehensive Table Key *(continued)*

E. Protein Identification and Characterization

1. Mass Spectrometry

Comprehensive Table Key *(continued)*

Comprehensive Table Key *(continued)*

Comprehensive Table Key (*continued*)

Comprehensive Table Key *(continued)*

Nomenclature Guide

We would like to thank Maarten Hoek, Cold Spring Harbor Laboratory for updating this guide, which was originally compiled by Elaine Cantwell, Cold Spring Harbor Laboratory Press.

This guide is by no means a definitive or exhaustive list. In most cases, gene and allele names and symbols are shown in italics (mouse and rat gene names are an exception), but names of proteins are always shown in Roman type.

Arabidopsis

Arabidopsis *Genes*

- Names may be all caps (if wild type) or all lowercase (if mutant). Thus, *ALPHABETICA* denotes the wild type and *alphabetica* denotes the mutant strain of that wild type. Gene symbols (abbreviated version of the gene name) follow the same rule: all caps for wild type: all lowercase for mutant. Symbols are shown in italics: *ABC*, *abc*. Different genes having the same symbol are distinguished by different numbers. The numbers are part of the gene symbol and should be closed up and italicized (e.g., *abc4* and *abc5*).

- Phenotypes may be designated by the gene symbol in Roman type, with the first letter capitalized, followed by a plus or a minus sign. Thus, Abc$^+$ describes the wild type and Abc$^-$ the mutant. The (+/−) should be superscripted.

- Different alleles of the same gene are distinguished by a hyphen between the name and the following number (e.g., *abc4-1* and *abc5-2*). Note that the hyphen is part of the allele name and should be italicized. Do not superscript, change the hyphen to an en dash, or insert thin spaces into the name.

- For *Arabidopsis*, there is no difference in nomenclature between dominant and recessive genes. Occasionally, a *D* may be added to the end of an allele name (*abc5-2D*); this indicates that the allele is dominant to the wild type.

Arabidopsis *Proteins*

- Proteins are written in all caps, without italics (e.g., ABC).

Arabidopsis *Chromosomes*

- The five chromosomes are written in Arabic numbers as Chromosomes 1, 2, 3, 4, and 5. It is acceptable to abbreviate them as chr.1, chr.2, etc.

Caenorhabditis

Caenorhabditis elegans *Genes*

- Gene names consist of three letters, a hyphen, and an Arabic number. All should be italicized. The name may be followed by an italicized Roman numeral to indicate the linkage group on which the gene maps. Examples are *dpy-5*, *let-37*, *dpy-5 I*, and *let-37 X*. Note the space before the Roman numeral, although all characters are italic.

- The wild-type allele of a gene is designated by a plus sign immediately after the gene name; this can be inside or outside parentheses, for example, either *dpy-5(+)* or *dpy-5⁺* . Either form is acceptable, as long as it is used consistently throughout the manuscript.

- When gene and mutation names are used together, the mutation name is attached to the gene name in parentheses. The complete (composite) name should be italicized, as in *dpy-5(e61)*.

- Phenotype characteristics may be described in words (dumpy, uncoordinated). However, a nonitalicized three-letter abbreviation that corresponds to a gene name may also be used, and in these, the first letter of the abbreviation is capitalized (e.g., Dpy for dumpy; Unc for uncoordinated).

- Transposons are called Tc1, Tc2, Tc3, etc. They are not italicized except when included in a genotype or as an insertion by means of addition following a double colon, for example, *unc-54(r293::Tc2)*.

Caenorhabditis *Proteins*

- Protein products are indicated by the relevant gene name written in Roman caps, as in UNC-13.

Caenorhabditis *Strains*

- Strains are given nonitalicized names consisting of two uppercase letters followed by a number. The strain letters refer to the laboratory of origin. Examples are CB1833 and MT688.

Caenorhabditis *Chromosomes*

- *Caenorhabditis* chromosomes are italicized. There are five pairs of autosomal chromosomes (Chromosomes *I*, *II*, *III*, *IV*, and *V*) plus one sex chromosome (*X*, which refers to the letter X and not the Roman numeral ten). Some authors may refer to them as linkage groups; thus, you may also see *LGI*, *LGII*, *LGIII*, *LGIV*, *LGV*, and *LGX*. Either format is correct.

Danio

Zebrafish (Danio rerio and Other Species) Genes

- Gene names are all lowercase italic. Hyphens or other punctuation is rarely seen. Gene symbols are all lowercase italicized letters or lowercase letters followed by an italicized number (plus occasionally another letter). Examples include *engrailed*, *engrailed2b*, *no tail*, *eng*, *eng2b*, and *ntl*.
- Wild-type and mutant alleles are indicated by superscripted plus (for wild type) or mutant allele designations. Superscripts, either alone or in combinations, may refer to the laboratory of origin or to dominance (letter *d* in first position). Examples would be cyc^+, cyc^{m101}, and lof^{dt2}.
- Phenotypes are designated by the name or symbol of the mutation in italics (*cyclops*, *cyc*).
- Zebrafish strains are nonitalic and uppercase (AB, WIK, DAR).

Zebrafish Proteins

- The protein symbol is the same as the gene symbol, but in Roman type and with the first letter uppercase (Eng, Eng2b, Ntl).

Zebrafish Chromosomes

- There are 25 zebrafish chromosomes, designated by Arabic numbers 1–25.

Dictyostelium

Dictyostelium Genes

- Gene names are composed of three italic lowercase letters followed by a capital (*rdeA*, *tagB*, etc.).
- Mutant alleles follow this format but also include the mutant in parentheses, closed up: *rdeA(AK235)*.
- Phenotypes are not italicized and include a superscript letter (e.g., neo^r).

Dictyostelium *Proteins*

- The protein product follows the same format as the respective gene, but the first letter is capitalized and the protein name is Roman. Thus, the protein product of gene *regA* is written as RegA. The mutant form would be shown as RegA(D212A).

Dictyostelium *Chromosomes*

- The *Dictyostelium* chromosomes are numbered 1–6 (Arabic numbers).

Drosophila

Drosophila *Genes*

- The gene name begins with an uppercase letter if named for a mutant phenotype that is dominant to the wild type, or if named after a protein product or other molecular feature. A gene named after a mutant phenotype that is recessive to the wild type is shown in all lowercase letters.
- When a full gene name or gene symbol is used to indicate phenotype, rather than genotype, it is not italicized. Therefore *white* indicates a genotype but white indicates a phenotype.
- There are also mitochondrial genes (having the prefix *mt*) and genes named for RNAs (beginning with either *tRNA* or *snRNA*), for species other than *D. melanogaster* (such as *Dsim\y* to indicate the yellow gene in *Drosophila simulans*), or for activity or modification (*enhancer, lethal, sterile, suppressor*). Prefixes and symbols for these names are italicized just as are the full names.
- Alleles at a particular gene are designated by the same name and symbol but are differentiated by distinguishing superscripts. Some authors may choose to separate with a hyphen (e.g., *white-apricot*). All plus or minus symbols within an allele superscript should be italicized.
- Transposable elements should be italicized (e.g., *P* element, *cis* element), and their symbols are included as a prefix in symbols of the genes they carry (e.g., *H\T* for *H*-element transposase). Use a hyphen if the element is being used as part of an adjective ("*P*-element transcription") but a space if it is used as a noun ("the effect of the *P* element in...").

Drosophila *Proteins*

- Protein products named for a gene may be indicated by that gene's symbol, but in all-cap Roman letters. When the full gene name is used for the protein (rather than the gene symbol), only the first letter of the protein name is capitalized. Just to make things a bit more complicated, when the gene name or symbol is used as an adjective modifying the word "protein," the rules for gene names apply. Therefore, the protein product of the *hedgehog* gene could correctly be described in four different ways: as the *hedgehog* protein, as the *hh* protein, as Hedgehog, or as HH.

- There are no standard rules for denoting proteins not named for a gene. They may be all-cap symbols (XDH), mixed-case symbols (AchE), or described as a product of a fully named gene (product of the *rosy* gene).

Drosophila *Chromosomes*

- *Drosophila* chromosomes are italicized. They may be shown as capital letters (*X*, *Y*) or Arabic numbers (*2, 3, 4*).

- In *Drosophila*, the letter *X* and the number *1* are accepted as synonymous; however, the number *1* is preferred in formal descriptions.

Gallus gallus

Chicken and Other Poultry Genes

- Gene symbols are designated by uppercase italic letters or by a combination of uppercase letters and Arabic numbers (e.g., *PO, MM7, GPDA*). No superscripts, subscripts, Roman numerals, or Greek letters should be used. Gene symbols are generally italicized.

- Allele designations include only capital letters or Arabic numbers and are written on the same line as the gene symbol (no superscript!). Allele characters are separated from gene characters by an asterisk. The entire item should be closed up and italicized (e.g., *OV*A, OV*B, EAA*7, EAA*1*).

- Front slashes or horizontal lines may separate alleles and indicate chromosome location. Hence, all of these forms are correct: *GHR*A/GHR*B*, *GHR*A/*B*, and $\dfrac{GHR*A}{GHR*B}$.

- Loci on the same chromosome but phase unknown are separated by a comma (*GHR*A,GHR*B*). Note that neither the comma nor the front slash in the bulleted section directly above is italicized.

Poultry Proteins

- This information was not available at press time.

Poultry chromosomes

- Poultry sex chromosomes are W and Z (Roman, uppercase). Autosomes are designated by Arabic numbers.

Homo sapiens

Human Genes

- Gene symbols are designated by uppercase Roman letters or a combination of uppercase letters and Arabic numbers. No subscripts or superscripts are used. No Roman numerals or Greek symbols are used; change either of these to their Arabic equivalents.

Human Proteins

- Proteins may contain uppercase Roman letters by themselves or in combination with Arabic numbers. Hyphens may be used. Examples include TPO-R, ADAM17, and L1CAM.

Human Chromosomes

- Sex chromosomes are X, Y, or 0 (for unknown assignment); all are written in uppercase Roman type. Autosomes are given Arabic numerals: 2, 4, 5, 13, 22, etc.

Mus and Rattus

Mouse (Mus musculus and other species) and Rat (Rattus norvegicus, et al.) genes

- The names of genes (such as recessive spotting or Purkinje cell degeneration) are not italicized, although the symbols are. All gene names, regardless of whether they are dominant or recessive, begin with a lowercase letter except when uppercase is required because of a proper name, for example in Purkinje cell degeneration.

- Rodent genes and loci are given short symbols (mixtures of Roman letters and Arabic numbers, with no punctuation except a hyphen in certain circumstances) as abbreviations. The initial letter is normally uppercase, and the rest is lowercase. However, if the gene is identified only by a recessive mutant phenotype, the symbol begins with a lowercase letter. Symbols for dominant or semidominant phenotypes begin with an uppercase letter (e.g., recessive spotting, *rs*; colboma, *Cm*; Purkinje cell degeneration, *Pcd)*. Allelic mutations at the same locus may be given a laboratory code as a superscript. Thus, *Pcd³ʲ* designates the third new allele of *Purkinje cell degeneration,* identified by The Jackson Laboratory.

- Mutations of structural genes are given a superscript m#Labcode (e.g., *Mod1aᵐ¹ᴸʷˢ*). Note that everything remains italicized.

Mouse and Rat Proteins

- Proteins in mouse and rat may be descriptive words in lowercase Roman (coiled-coil, beta-defensin) or all uppercase abbreviations (HPC-1, PCD).

Mouse and Rat Chromosomes

- Chromosomes for mouse and rat are shown in Arabic numbers 1–20, plus the letter X. When referring to a particular chromosome, the c is uppercase: Chromosome 5, or Chr 5. Note that there should be no period after Chr when using the abbreviated word.

Oryza (Rice)

Oryza sativa *Genes*

- Rice gene names are written in Roman type, with the initial letter uppercase. All gene names are followed by a hyphen and a locus designator that is generally an Arabic number. Rice gene symbols may consist of three to five letters, written in italics, followed by the locus designator, closed up and in Roman type (e.g., *Glh*2, *Pi*12, *Piz*, *Pi*ta). Different alleles of the same gene are distiguished by adding a numerical suffix, separated by a dash or hyphen, to the gene full name or the gene symbol (e.g., *Pgi*1-1, *Pgi*1-2).

Oryza sativa *Proteins*

- Protein symbols follow the respective gene symbol but are all caps (e.g., TNP2). There should be no space or hyphen.

Oryza sativa Chromosomes

- Rice chromosomes are indicated by Arabic numerals (1–12).

Saccharomyces

Yeast (Saccharomyces cerevisiae) *Genes*

- Gene names consist of three letters (the gene symbol) followed by an Arabic number (e.g., *ADE12*), closed up and italicized.
- Dominant alleles (most often wild type) are shown by all uppercase letters; recessive alleles are shown by all lowercase.

Yeast Proteins

- Yeast proteins are indicated by the gene symbol in Roman type, followed by a "p" suffix (e.g., Ade5p, Cdc28p, Cup1p, Spc1205p). However, the "p" suffix is often omitted in publications.
- Phenotypes are designated by the gene symbol in Roman type, with the first letter uppercase. Wild-type or mutant status is indicated by a superscript plus or minus sign, respectively (e.g., Arg$^-$ for a strain requiring arginine).

Yeast Chromosomes

- *Saccharomyces cerevisiae* has 16 chromosomes, which are shown by Roman numerals I–XVI.
- *Saccharomyces pombe* has only three chromosomes, which are shown by Roman numerals: Chromosome I, Chromosome II, and Chromosome III.

Xenopus

Xenopus laevis *(Frog) Genes*

- *Xenopus* gene names are officially not italicized, although they often are in journals.

Xenopus *Proteins*

- Proteins may be lowercase Roman words (sonic hedgehog, zinc finger) or all uppercase symbols (SH, ZIC1). Most *Xenopus* proteins contain a number, which should be closed up to the symbol (SMAD3, PDCD9).

Xenopus *Chromosomes*

- There are 18 chromosomes, designated by the Arabic numbers 1–18.

Zea mays

Zea mays *(Maize)* Genes

- The name and symbol of the gene or locus should be represented with lowercase, italic characters (*defective kernel12, dek12*). Note that no hyphen separates the gene name from a numerical suffix, which is a change from previous usage.

- You may also see examples of the homozygous genotypes with two or more unlinked genes, the genes being separated by semicolons, e.g., *a1;a2;c1;c2;r*. Note that these are closed up. However, if linked, the genes are separated by spaces, e.g., *C1 sh1 bz1 Wx1*. Heterozygous genotypes are written with a slash separating the sets of linked genes, e.g., *C1 Bz1/c1 bz1*, with spaces between; but if the genes are unlinked, the proper designation is *Sh2/sh2; Bt2/bt2*. Do not change the format in cases like this; leave as it has been provided by the author.

- Alleles may have the same designation as the gene (dek12), but if dominant, may be indicated with an initial cap (Dek12). Mutant alleles may contain ending hyphens (Dek12-); these are not minus signs and are not superscripted.

Zea *Proteins*

- Protein products have the same name as the gene or allele, but Roman and in all caps (e.g., ADH1); however, if the name is written out in full (alcohol dehydrogenase), it should be all lowercase.

Zea *Chromosomes*

- Chromosomes are indicated by Arabic numerals (1–10). Some maize stocks contain an additional chromosome termed the B chromosome.

At-a-Glance Quick Reference Guide to Species Nomenclature

See the detailed Nomenclature Guide for any items that are not listed here or for any sections that contain the note "See special details."

Species	Gene names (always italic except for *Mus*)	Gene symbols (always italic)	Proteins (always Roman)	Chromosomes (always Roman except as noted)
Arabidopsis	Can be all cap or all lowercase	All cap or all lowercase. Italicize/close up any numbers; + or – are ital/superscript.	All caps. No italics.	1, 2, 3, 4, and 5. Okay to abbreviate as chr. 1, chr. 2, etc.
Caenorhabditis elegans	Three letters, hyphen, and number. May contain a Roman numeral.	See gene names.	Same as gene name, but written in all caps and Roman.	*I, II, III, IV, V,* and *X.* All are italicized.
Chicken (*Gallus gallus* and other fowl)	Same as gene symbols. See special details.	Uppercase letters, or uppercase letters plus Arabic numbers.	Information to come.	W and Z, or Arabic numbers.
Danio rerio (zebrafish)	Word or abbreviations, all lowercase. See special details.	All lowercase words or letters; may be with number. Close up all.	Same as gene symbol, but with first letter uppercase.	Arabic numbers 1–25.
Dictyostelium	Three lowercase letters followed by a cap letter. Close up all.	See gene names; see special details.	Same as gene name, but Roman type and cap first letter.	1, 2, 3, 4, 5, and 6.
Drosophila	Can begin with either uppercase or lowercase letter.	See special details about phenotypes. Important.	May be written in several different ways, all correct. See special details.	May be X or Y, or Arabic numbers. All are italicized.

Species				
Homo sapiens	Uppercase letters, or uppercase letters plus Arabic numbers.	Do not subscript or superscript. No Roman numerals or Greek characters.	Only uppercase letters or combo of uppercase plus Arabic numbers. Hyphens are okay.	X, Y, 2, 4, 5, 13, 22, etc.
Mus and *Rattus* species (mouse and rat)	Not italicized; begins with lowercase letter unless first word is a proper name.	May begin with uppercase or lowercase letter. May not match case with its "word" name.	May be either descriptive word or all uppercase abbreviation.	Letter X, plus numbers 1–20. Okay to abbreviate as Chr #. No period after Chr.
Oryza (rice)	Roman type; initial letter lowercase.	Three or four lowercase letters; may end with uppercase letter.	Same as gene symbol, but all uppercase. No space, no hyphens.	Information to come.
Saccharomyces and other yeast species	Three letters plus Arabic number. Close up.	See gene names.	Same as gene name. Initial letter uppercase.	*S. cerevisiae*, Roman numerals I–XVI. *S. pombe*, I–III.
Xenopus laevis and other frog species	May be lowercase word or all uppercase letters with or without Arabic number. Hyphen is okay.	See gene names.	Lowercase word or all uppercase symbols. Most will contain a number. No hyphen. Close up all.	Numbers 1–18.
Zea mays (maize)	Lowercase letters plus numbers, except for alleles that have uppercase first letter. Close up. No hyphens.	See gene names.	Same as gene name, but all caps.	Numbers 1–10, and B.

Useful World Wide Web Sites

All World Wide Web addresses are accurate to the best of our knowledge at the time of printing. Many of these sites provide multiple useful links.

Resource	URL	Type of Information
Organizations		
American Society for Cell Biology	http://www.ascb.org	Educational resource
Harvard University Biosciences Index	http://mcb.harvard.edu/Biolinks.html	Links to multiple scientific sites
HudsonAlpha Genome Sequencing Center (formerly the Stanford Human Genome Center)	http://hagsc.org	Genomic information for human and *Xenopus*
Indiana University	http://iubio.bio.indiana.edu	IUBio archive for biology data and software
Microscope Society of America	http://msa.microscopy.com	Information and links for all microscopical imaging, analysis, and diffraction techniques
Molecular Expressions	http://micro.magnet.fsu.edu/index.html	Photo galleries explore optical microscopy
Purdue University	http://www.cyto.purdue.edu	Flow cytometry. Includes cytometry discussion board
Wellcome Trust Sanger Institute	http://www.sanger.ac.uk	Large-scale genome research, sequencing, and analysis
Whitehead Institute	http://www.wi.mit.edu	Research topics
Sources for Materials		
American Type Culture Collection	http://www.atcc.org	Strains and cell lines
Antibody Resource	http://antibodyresource.com	Antibody suppliers, contract services, educational resources for immunology/biotechnology, and databases"
BioSupplyNet	http://www.biosupplynet.com	Searches for scientific supplies and suppliers
Cold Spring Harbor Laboratory Press	http://www.cshlpress.com	Scientific titles, laboratory manuals, and journals
Compendium of Cre, floxed, and Cre excision reporter lines	http://www.mshri.on.ca/nagy	Conditional genetic mice

Comprehensive Protocol Collection, Ambrose Lab, Dartmouth	http://wormbase.org	Protocols for *C. elegans* experimentation
Coriell Cell Repositories	http://ccr.coriell/org	Collections of cell cultures and DNA derived from cell cultures
Cybergenome Technologies	http://cybergenome.com/index.html	Searches for genome products
EMMA	http://www.emmanet.org	European mouse mutant archive
ExPASy	http://www.expasy.com	Multifaceted bioinformatics and proteomics company
Frimorfo	http://www.frimorfo.com	Company specializing in histological characterization of genetically modified rodents
IGTC	http://www.genetrap.org	International gene trap consortium. Resource for embryonic stem cells with gene trap insertions in all genes in the mouse genome
IMR	http://www.jax.org/imr/index.html	Induced mutant mouse resource
MAFF DNA Bank	http://www.dna.affrc.go.jp	Distributes cDNAs, RFLP markers and YAC clones obtained in rice and animal genome projects
MMHCC	http://emice.nci.nih.gov/emice/mouse_models	NCI mouse models of human cancers consortium
MMRRC	http://www.nih.gov/science/models/mouse/resources/mmrrc.html	NIH mutant mouse regional resource centers
Xenopus Express	http://www.xenopus.com	Specimen source
Databases		
5S Ribosomal RNA	http://www.man.poznan.pl/5SData	5S ribosomal RNA data bank

(continued on following pages)

Resource	URL	Type of Information
Databases (*continued*)		
CEPH	http://cephb.fr/en/cephdb	Genotypes
Codon use	http://www.kazusa.or.jp/codon	DNA motifs and patterns
COGs	http://www.ncbi.nlm.nih.gov/COG	Orthology/phylogenetic profiles
dbEST from NCBI	ftp://ftp.ncbi.nih.gov/blast/db (click est_human.Z, est_mouseZ, or est_others.Z)	Expressed tag sequences
DDBJ	http://www.ddbj.nig.ac.jp	DNA sequences
DIP	http://dip.doe-mbi.ucla.edu	Protein interactions
EBI (EMBL)	http://www.ebi.ac.uk	DNA sequences
EPD	http://www.epd.isb-sib.ch	Eukaryotic promoter database
FlyBase	http://flybase.org	Species-specific for *D. melanogaster*
GenPept	ftp://ftp.ncbi.nih.gov/genbank (click genpept.fsa.qz)	Protein translation of GenBank
Gobase	http://gobase.bcm.umontreal.ca	Mitrochondrial sequences
INTACT	http://www.ebi.ac.uk/intact/index.jsp	Protein interactions
INTERACT	http://bioinf.manchester.ac.uk/interactpr.htm	Protein interactions
MIPS	http://mips.helmholtz-muenchen.de/proj yeast/CYGD	Yeast protein interactions
MSDS databases	http://www.msdssearch.com	Materials safety datasheet information
NCBI (GenBank)	http://www.ncbi.nlm.nih.gov	DNA sequences
OWL	ftp://ftp.ncbi.nih.gov/repository/OWL (click owl.fasta.Z)	Nonredundant composite protein sequence database produced from SwissProt, PIR 1-3, Genabank translated, and Brookhaven databases

PDB	http://www.rcsb.org/pdb	Three-dimensional macromolecular structure data determined mainly by X-ray crystallography and nuclear magnetic resonance
PIR	http://pir.georgetown.edu/pirwww	Protein sequences
PlantCARE	http://bioinformatics.psb.ugent.be/webtools/plantcare/html	Plant *cis*-acting regulatory elements
REBASE	http://rebase.neb.com	Restriction enzymes
Ribosomal RNA mutations	http://ribosome.fandm.edu	Ribosomal RNA mutations
RNA modification	http://medlib.med.utah.edu/RNAmods	RNA modifications
SwissProt	http://www.expasy.ch	Protein sequences
The Genome Database	http://gdwww.gdb.org	Official worldwide database for the annotation of the human genome
tRNA Genes	ftp://ftp.ebi.ac.uk/pub/databases/plmitrna	Higher-plant mitrochondria
TRRD	http://www.epd.isb-sib.ch/promoter_elements	Transcriptional regulatory regions
United States National Library of Medicine	http://www.nlm.nih.gov	Searches of worldwide scientific and medical journals
WormPD	http://www.proteome.com	Species-specific database for *C. elegans*
Yeast splice sites by M. Ares, Jr. laboratory	http://www.cse.ucsc.edu/research/compbio/yeast_introns.html	Yeast splice sites
YPD	http://www.proteome.com	Species-specific database for *S. cerevisiae*
Gene Annotation		
Gene Ontology Consortium	http://www.geneontology.org	Structured, controlled vocabulary describing gene product processes, functions, and localizations

(continued on following pages)

Resource	URL	Type of Information
Gene Annotation (*continued*)		
GeneCards	http://bioinformatics.weizmann.ac.il/cards	Functions of human genes and their roles in disease
GeneMap99	http://www.ncbi.nlm.nih.gov/genemap	Mapping information for human genes
LocusLink	http://www.ncbi.nlm.nih.gov/LocusLink	Curated sequences and information on human, mouse, fruit fly, rat, and zebrafish genes
Mouse Genome Informatics	http://www.informatics.jax.org	Annotation of mouse genes and sequences
Proteome	http://www.proteome.com	Annotation of proteins from *S. cerevisiae*, *Schizosaccharomyces pombe*, and *C. elegans*
Saccharomyces Genome Database	http://www.yeastgenome.org	Annotated information about *S. cerevisiae* genes and sequences
SOURCE	http://genome-www.stanford.edu/source	Compiled information about genes and cDNAs for human, mouse, and rat"
TAIR	http://www.arabidopsis.org	Biological information about *Arabidopsis thaliana* genes
Unigene	http://www.ncbi.nlm.nih.gov/UniGene	Groups expressed sequence tags into unique human, mouse, rat, cow, and zebrafish gene-oriented clusters
WormBase	http://www.wormbase.org	Biological information about *C. elegans*
Sequence Submission		
BankIt and Sequin	http://www.ncbi.nlm.nih.gov	Web-based sequence submission tools for Genbank, EMBL, and DDJB
Webin	http://www.ebi.ac.uk/Tools/index.html	Web-based interface preferred for EMBL submission

Sequence Alignment Software

BLAST	http://blast.ncbi.nlm.nih.gov/Blast.cgi	Local alignment of sequences based on word "seeds" and graphical representation of alignment
FASTA	http://fasta.bioch.virginia.edu	Single, local (Smith-Waterman) alignment between two sequences
FASTX/FASTY	http://fasta.bioch.virginia.edu	Compares translated DNA sequences to protein sequence database
LALIGN, LALIGN0, PLALIGN	http://fasta.bioch.virginia.edu	Compares two DNA or two protein sequences to identify regions of sequence similarity
LFASTA	http://fasta.bioch.virginia.edu	Local sequence alignment
PRSS	http://fasta.bioch.virginia.edu	Evaluates the significance of pairwise similarity scores of two DNA or two protein sequences
SIM	http://expasy.ch/tools/sim-prot.html	Local alignments in order of similarity score

Scoring Matrices

DNA PAM matrices	http://blast.advbiocomp.com	Predicts evolutionary distance among DNA sequences

DNA: Multiple Sequence Alignment Software

CLUSTAL	http://www.clustal.org	Shows sequences aligned in blocks
DIALIGN	http://bibiserv.techfak.uni-bielefeld.de/dealign	Sequence tree showing degrees of similarity
HMMER	http://hmmer.janelia.org	Uses hidden Markov models to produce multiple sequence alignment of DNA, RNA, or protein sequences
MACAW	ftp://ftp.ncbi.nih.gov/pub/macaw	Multiple sequence alignment based on blocks

(continued on following pages)

Resource	URL	Type of Information
DNA: Multiple Sequence Alignment Software (*continued*)		
MSA	http://www.psc.edu/general/software/packages/msa/msa.html	Initally shows heuristic alignment based on a progressive pairwise alignment
MultiAlin	http://multalin.toulouse.inra.fr/multalin	Output as colored image, plain text, or HTML
Parallel PRRN	htp://www.genome.jp/tools/prrn/prrn_help.html	Progressive global alignment, randomly or doubly nested
PIMA	http://searchlauncher.bcm.tmc.edu/multi-align/Options/pima.html	Multiple sequence alignment
SAM	http://compbio.soe.ucsc.edu/sam.html	Hidden-Markov-model–based multiple sequence alignment; database search for new family members
Database Similarity Search Software		
AB-BLAST	http://blast.advbiocomp.com	Performs similarity searches of proteins and nucleotide sequence databases
BLAST2	http://www.ncbi.nlm.nih.gov/BLAST	newer version 2 of BLAST
BLAST-Genome Sequences	http://www.ncbi.nlm.nih.gov/genome/seq/BlastGen/BlastGen.cgi?taxid=9606	Searches a query sequence against the human genome database
FASTS/TFASTS, FASTF/TFASTF	http://fasta.bioch.virginia.edu	Compares sequence of peptide fragments (mass-spec) or an ordered peptide mixture against a protein database
PHI-BLAST	http://www.ncbi.nlm.nih.gov/BLAST	Finds protein sequences that share an amino acid pattern
PSI-BLAST	http://www.ncbi.nlm.nih.gov/BLAST	Searches for similar protein sequences using a multiple sequence alignment

DNA: Motifs and Patterns Search Software

BLIMPS-BLOCKS	http://blocks.fhcrc.org	Scores sequence against BLOCKS database or blocks against sequence
BLOCKS	http://blocks.fhcrc.org	Blocks are conserved patterns of amino acid sequences of the same length (no gaps) found in members of a protein family
Cytoscape	http://cytoscape.org/	Network visualization and analysis
emotif	http://motif.stanford.edu/distributions/emotif/index.html	Forms motifs or subsets of aligned sequences
Gibbs sampler	http://bayesweb.wadsworth.org/gibbs/gibbs.html	Locates conserved patterns in sequences
MAST	http://meme/sdsc.edu/meme/doc/mast.html	Searches databases for sequences that contain one or more motifs
MEME	http://meme.sdsc.edu/meme/cgi-bin/meme.cgi	Detects conserved sequence patterns of the same length; no gaps; no covariations between positions
Meta-MEME	http://metameme.sdsc.edu	Combines DNA or protein motif patterns from MEME into a hidden Markov model
Osprey	http://biodata.mshri.on.ca:80/osprey/servlet/Index	Network visualization and analysis
Pratt	http://expasy.org/tools/pratt	Profile/pattern identification

DNA: Genes, Exons, Introns Search Software

ECOCYC	http://ecocyc.org	Metabolic pathway analysis
Expression Profiler	http://www.ebi.ac.uk/expressionprofiler	Gene expression profile analysis
FGENES	http://genomic.sanger.ac.uk/gf/gf.html	Predicts genes and exons using pattern-based structure predictions

(continued on following pages)

Resource	URL	Type of Information
DNA: Genes, Exons, Introns Search Software (*continued*)		
GeneMark	http://exon.biology.gatech.edu	Identifies protein-coding regions in prokaryotic or eukaryotic genomic DNA sequences
Genie	http://www.fruitfly.org/seq_tools/genie.html	Finds potential genes in *Drosophila* genomic sequences
GENSCAN	http://genes.mit.edu/GENSCAN.html	Predicts genes in genomic DNA based on probabilistic models in gene structure in various organisms
STRING	http://string.embl-heidelberg.de	Gene colocalization
DNA: Promotors, Transcription-factor-binding Sites Search Software		
CorePromotor	http://rulai.cshl.edu/tools/genefinder/CPROMOTER/index.htm	Predicts transcriptional start sites and localizes them into 50–100-bp core-promotor regions
GenomeInspector	http://www.genomatix.de/online_help/help_regionminer/GenomeInspector.html	Detects distance correlation between open reading frames and transcription-binding sites on megabases of nucleotide sequences
NNPP	http://www.fruitfly.org/seq_tools/promotor.html	Finds eukaryotic and prokaryotic promoters in DNA sequence
TESS	http://www.cbil.upenn.edu/cgi-bin/tess/tess	Finds potential transcription-factor-binding sites
TFBIND	http://tfbind.ims.u-tokyo.ac.jp	Searches for transcription-factor-binding sites including TATA, GC, and CCAAT boxes, and transcription start sites
Transfac	http:/www.gene-regulation.com/pub/databases/transfac/doc/toc.html	A relational database of transcription-factor cis-acting binding sites and transcription factors from many organisms

DNA: Other Regulatory Sites Software

polyadq	http://rulai.cshl.edu/tools/polyadq/polyadq_form.html	Decides if given AATAA or ATTAAA hexamer is true poly(A) signal

RNA: Secondary Structure Software

MFOLD	http://bioweb.pasteur.fr/seqanal/interfaces/mfold.html	Predicts RNA secondary structure
Vienna RNA package	http://www.tbi.univie.ac.at/~ivo/RNA	Calculates predictions of RNA structure

RNA: RNA-specifying Genes, Motifs Search Software

tRNAscan-SE	http://lowelab.ucsc.edy/tRNAscan-SE	Searches for *t-RNA* genes in genomic DNA or RNA sequences

Proteins: Motifs, Patterns, Profiles Software

Pfam	http://www.sanger.ac.uk/Software/Pfam	Enables rapid and automatic classification of predicted proteins into protein domain families
PFSEARCH	http://www2.ebi.ac.uk/ppsearch	Scans sequence against patterns in PROSITE
PowerBLAST	http://www.ncbi.nlm.nih.gov/Kuehl/prefinished/powblast.html	Provides added functionality over traditional BLAST programs
PROSITE	http://ca.expasy.org/prosite	Scans sequence against patterns in PROSITE
ScanProsite	http://ca.expasy.org/tools/scnpsit3.html	Scans sequence against patterns in PROSITE or Swiss-Prot

NOTES

NOTES

NOTES

NOTES

NOTES

Cautions Appendix

General Cautions

This Cautions Appendix is not exhaustive or all inclusive. Please consult your local safety office and/or the manufacturer's safety guidelines for current and specific product information.

The following general cautions should always be observed.

- **Become completely familiar** with the properties of the substances used **before** beginning the procedure.

- **The absence of a warning** does not necessarily mean that the material is safe, since information may not always be complete or available.

- **If exposed to toxic substances,** contact your local safety office immediately for instructions.

- **Use proper disposal procedures** for all chemical, biological, and radioactive waste.

- **For specific guidelines on appropriate gloves,** consult your local safety office.

- **Handle concentrated acids and bases with great care.** Wear goggles and appropriate gloves. A face shield should be worn when handling large quantities. Do not mix strong acids with organic solvents because they may react. Sulfuric acid and nitric acid especially may react highly exothermically and cause fires and explosions. Do not mix strong bases with halogenated solvent because they may form reactive carbenes that can lead to explosions.

- **Handle and store pressurized gas containers** with caution as they may contain flammable, toxic, or corrosive gases; asphyxiants; or oxidizers. For proper procedures, consult the Material Safety Data Sheet that must be provided by your vendor.

- **Never pipette solutions using mouth suction.** This method is not sterile and can be dangerous. Always use a pipette aid or bulb.

219

- **Keep halogenated and nonhalogenated solvents separately** (e.g., mixing chloroform and acetone can cause unexpected reactions in the presence of bases). Halogenated solvents are organic solvents such as chloroform, dichloromethane, trichlorotrifluoroethane, and dichloroethane. Some nonhalogenated solvents are pentane, heptane, ethanol, methanol, benzene, toluene, *N,N*-dimethylformamide (DMF), dimethyl sulfoxide (DMSO), and acetonitrile.

- **Laser radiation,** visible or invisible, can cause severe damage to the eyes and skin. Take proper precautions to prevent exposure to direct and reflected beams. Always follow manufacturers' safety guidelines and consult your local safety office. See flash lamps caution below for more detailed information.

- **Flash lamps,** due to their light intensity, can be harmful to the eyes. They also may explode on occasion. Wear appropriate eye protection and follow the manufacturers' guidelines.

- **Photographic fixatives and developers** also contain chemicals that can be harmful. Handle them with care and follow manufacturers' directions.

- **Power supplies and electrophoresis equipment** pose serious fire hazards and electrical shock hazards if not used properly.

- **Microwave ovens and autoclaves** in the lab require certain precautions. Accidents have occurred involving their use (e.g., when melting agar or bacto-agar stored in bottles or sterilizing). If the screw top is not completely removed and insufficient space is available for the steam to vent, the bottles can explode and cause severe injury when the containers are removed from the microwave or autoclave. Always completely remove bottle caps before microwaving or autoclaving. An alternative method for routine agarose gels that do not require sterile agar is to weigh out the agar and place the solution in a flask.

- **Ultrasonicators** use high-frequency sound waves (16–100 kHz) for cell disruption and other purposes. This "ultrasound," conducted through air, does not pose a direct hazard to humans, but the associated high volumes of audible sound can cause a variety of effects including headache, nausea, and tinnitus. Direct contact of the body with high-intensity ultrasound (not medical imaging equipment) should be avoided. Use appropriate ear protection and display signs on the door(s) of laboratories in which the units are used

- **Use extreme caution when handling cutting devices** such as microtome blades, scalpels, razor blades, or needles. Microtome blades are extremely sharp! Use care when sectioning. If you are unfamiliar with their use, have someone demonstrate proper procedures. For proper disposal, use the "sharps" disposal container in your lab. Discard used needles *unshielded*, with the syringe still attached. This prevents injuries (and possible infections) when manipulating used needles, since many accidents occur while trying to replace the needle shield. Injuries may also be caused by broken pasteur pipettes, cover slips, or slides.

General Properties of Common Chemicals

The hazardous materials list can be summarized in the following categories:

- Inorganic acids such as hydrochloric, sulfuric, nitric, or phosphoric are colorless liquids with stinging vapors. Avoid spills on skin or clothing. Spills should be diluted with large amounts of water. The concentrated forms of these acids can destroy paper, textiles, and skin, as well as cause serious injury to the eyes.

- Inorganic bases such as sodium hydroxide are white solids that dissolve in water and under heat development. Concentrated solutions will slowly dissolve skin and even fingernails.

- Salts of heavy metals are usually colored powdered solids that dissolve in water. Many are potent enzyme inhibitors and therefore toxic to humans and to the environment (e.g., fish and algae).

- Most organic solvents are flammable volatile liquids. Avoid breathing the vapors, which can cause nausea or dizziness. Avoid skin contact.

- Other organic compounds, including organosulfur compounds such as mercaptoethanol or organic amines, can have very unpleasant odors. Others are highly reactive and should be handled with appropriate care.

- If improperly handled, dyes and their solutions can stain not only your sample, but also your skin and clothing. Some of them are also mutagenic (e.g., ethidium bromide), carcinogenic, and toxic.

- Nearly all names ending with "ase" (e.g., catalase, β-glucuronidase, or Zymolase) refer to enzymes. There are also other enzymes with nonsystematic names such as pepsin. Many are provided by manufacturers in preparations containing buffering substances, etc. Be aware of the individual properties of materials contained in these substances.

- Toxic compounds are often used to manipulate cells. They can be dangerous and should be handled appropriately.

- Be aware that the toxicological properties of several of the compounds listed have not been thoroughly studied. Handle each chemical with the appropriate respect. Although the toxic effects of a compound can be quantified (e.g., LD_{50} values), this is not possible for carcinogens or mutagens for which one single exposure can have an effect. Realize that dangers related to a given compound may also depend on its physical state (fine powder vs. large crystals/diethylether vs. glycerol/dry ice vs. carbon dioxide under pressure in a gas bomb). Anticipate under which circumstances during an experiment exposure is most likely to occur and how best to protect yourself and your environment.

- All chemicals must be reagent grade or molecular biology grade, and the water used in the preparation of all solutions must be the highest quality available. Use sterile, glass-distilled, deionized water whenever possible. Unless otherwise stated, most solutions require sterilization either by filtration or autoclaving.

- Where the directions call for filter sterilization, solutions should be passed through a 0.22-μm filter before storing at the recommended temperature. A number of commercially available filters are suitable for syringe, large-scale, or bottle-top filtration. Most solutions can be stored for at least 6 months at room temperature, unless otherwise specified. Be sure that repeated use of a common stock of these solutions does not result in contamination. Aliquots will help to avoid this problem.

Hazardous Materials

Note: In general, proprietary materials are not listed here. Kits and other commercial items as well as most anesthetics, dyes, fixatives, and stains are also not included. Anesthetics also require special care. Follow the manufacturer's safety guidelines that accompany these products.Acetic acid (concentrated) must be handled with great care. It may be harmful by inhalation, ingestion, or skin absorption. Wear appropriate gloves and goggles and use in a chemical fume hood.

Acetic acid (glacial) is highly corrosive and must be handled with great care. It may be a carcinogen. Liquid and mist cause severe burns to all body tissues. It may be harmful by inhalation, ingestion, or skin absorption. Wear appropriate gloves and goggles and use in a chemical fume hood. Keep away from heat, sparks, and open flame.

Acetic anhydride is extremely destructive to the skin, eyes, mucous membranes, and the upper respiratory tract. It may be harmful by inhalation, ingestion, or skin absorption. Wear appropriate gloves and safety glasses and use in a chemical fume hood.

Acetone causes eye and skin irritation and is irritating to mucous membranes and the upper respiratory tract. Do not breathe the vapors. It is also extremely flammable. Wear appropriate gloves and safety glasses.

Acetonitrile (methyl cyanide) is very volatile and extremely flammable. It is an irritant and a chemical asphyxiant that can exert its effects by inhalation, ingestion, or skin absorption. Treat cases of severe exposure as cyanide poisoning. Wear appropriate gloves and safety glasses and use only in a chemical fume hood. Keep away from heat, sparks, and open flame.

Acrylamide (unpolymerized) is a potent neurotoxin and is absorbed through the skin (the effects are cumulative). Avoid breathing the dust. Wear appropriate gloves and a face mask when weighing powdered acrylamide and methylene-bisacrylamide. Use in a chemical fume hood. Polyacrylamide is considered to be nontoxic, but it should be handled with care because it might contain small quantities of unpolymerized acrylamide.

Actinomycin D is a teratogen and a carcinogen. It is highly toxic and may be fatal if inhaled, ingested, or absorbed through the skin. It may also cause irritation. Avoid breathing the dust. Wear appropriate gloves and safety glasses, and always use in a chemical fume hood. Solutions of actinomycin D are light sensitive.

Aldehyde, *see* **Formaldehyde**

Ammonium bicarbonate, NH_4HCO_3, may be harmful by inhalation, ingestion, or skin absorption. Wear appropriate gloves and safety glasses and use in a chemical fume hood.

Ammonium persulfate, $(NH_4)_2S_2O_8$, is extremely destructive to tissue of the mucous membranes and upper respiratory tract, eyes, and skin. Inhalation may be fatal. Wear appropriate gloves, safety glasses, and protective clothing. Use only in a chemical fume hood. Wash thoroughly after handling.

Ammonium sulfate, $(NH4)_2SO_4$, may be harmful by inhalation, ingestion, or skin absorption. Wear appropriate gloves and safety glasses.

Ampicillin may be harmful by inhalation, ingestion, or skin absorption. Wear appropriate gloves and safety glasses and use in a chemical fume hood.

Aprotinin may be harmful by inhalation, ingestion, or skin absorption. It may also cause allergic reactions. Exposure may cause gastrointestinal effects, muscle pain, blood pressure changes, or bronchospasm. Wear appropriate gloves and safety glasse and use only in a chemical fume hood. Do not breathe the dust.

Bacterial strains (shipping of): The Department of Health, Education, and Welfare (HEW) has classified various bacteria into different categories with regard to shipping requirements. Nonpathogenic strains of *E. coli* (such as K12) and *B. subtilis* are in Class 1 and are considered to present no or minimal hazard under normal shipping conditions. However, *Salmonella, Haemophilus,* and certain strains of *Streptomyces* and *Pseudomonas* are in Class 2. Class 2 bacteria are "Agents of ordinary potential hazard: agents which produce disease of varying degrees of severity...but which are contained by ordinary laboratory techniques." See the handbook edited by J.Y. Richmond and R.W. McKinney. 1999. *Biosafety in microbiological and biomedical laboratories* (BMBL), 4th edition. U.S. Department of Health and Human Services, Centers for Disease Control at http://www.cdc.gov/od/ohs/biosfty/biosfty.htm/.

BCIP, *see* **5-Bromo-4-chloro-3-indolyl-phosphate**

Benzyl alcohol is an irritant and may be harmful by inhalation, ingestion, or skin absorption. Wear appropriate gloves and safety glasses. Keep away from heat, sparks, and open flame.

Benzyl benzoate is an irritant and may be harmful by inhalation, ingestion, or skin absorption. Avoid contact with the eyes. Wear appropriate gloves and safety glasses.

5-Bromo-4-chloro-3-indolyl-phosphate (BCIP) is toxic and may be harmful by inhalation, ingestion, or skin absorption. Wear appropriate gloves and safety glasses. Do not breathe the dust.

Bromophenol blue may be harmful by inhalation, ingestion, or skin absorption. Wear appropriate gloves and safety glasses and use in a chemical fume hood.

CaCl₂, *see* **Calcium chloride**

Calcium chloride, CaCl₂, is hygroscopic and may cause cardiac disturbances. It may be harmful by inhalation, ingestion, or skin absorption. Do not breathe the dust. Wear appropriate gloves and safety glasses.

Calcium nitrate, Ca(NO₃)₂, is a strong oxidizer and reacts violently on contact with many organic substances. Handle with great care. It may be harmful by inhalation, ingestion, or skin absorption. Wear appropriate gloves and safety glasses. Keep away from heat, sparks, and open flame.

CAPS, *see* **3-(Cyclohexylamino)-1-propanesulfonic acid**

CBR, *see* **Coomassie brilliant blue**

Cesium chloride, CsCl, may be harmful by inhalation, ingestion, or skin absorption. Wear appropriate gloves and safety glasses.

Cetyltrimethylammonium bromide (CTAB) is toxic and an irritant and may be harmful by inhalation, ingestion, or skin absorption. Wear appropriate gloves and safety glasses. Avoid breathing the dust.

CHAPS, *see* **3-[(3-Cholamidopropyl)dimethyl-ammonio]-1-propanesulfonate**

CH₃CH₂OH, *see* **Ethanol**

CH₂Cl₂, *see* **Dichloromethane**

C₇H₇FO₂S, *see* **Phenylmethylsulfonyl fluoride**

Chicago Sky Blue is a possible mutagen and may be harmful by inhalation, ingestion, or skin absorption. Wear appropriate gloves and safety glasses and use in a chemical fume hood. Do not breathe the dust.

3-[(3-Cholamidopropyl)dimethyl-ammonio]-1-propanesulfonate (CHAPS) is an irritant and may be harmful by inhalation, ingestion, or skin absorption. Wear appropriate gloves and safety glasses.

Chymostatin is an irritant and may be harmful by inhalation, ingestion, or skin absorption. Do not breathe the dust.

Citric acid is an irritant and may be harmful by inhalation, ingestion, or skin absorption. It poses a risk of serious damage to the eyes. Wear appropriate gloves and safety goggles. Do not breathe the dust.

Coomassie brilliant blue (CBR) may be harmful by inhalation, ingestion, or skin absorption. Wear appropriate gloves and safety glasses.

Crystal Violet can cause severe burns. It may be harmful by inhalation, ingestion, and skin absorption. Wear appropriate gloves and safety goggles and use in a chemical fume hood, Do not breathe the dust.

CsCl, *seee* **Cesium chloride**

CTAB, *see* **Cetyltrimethylammonium bromide**

Cycloheximide may be fatal if inhaled, ingested, or absorbed through the skin. Wear appropriate gloves and safety glasses and use in a chemical fume hood.

3-(Cyclohexylamino)-1-propanesulfonic acid (CAPS) is an irritant and may be harmful by inhalation, ingestion, or skin absorption. Wear appropriate gloves and safety glasses. Do not breathe the dust.

DAB, *see* **3,3′-Diaminobenzidine tetrahydrochloride**

DCM, *see* **Dichloromethane**

Deoxycholate (DOC) may be harmful by inhalation, ingestion, or skin absorption. Do not breathe the dust. Wear appropriate gloves and safety glasses.

DEPC, *see* **Diethyl pyrocarbonate**

3,3′-Diaminobenzidine tetrahydrochloride (DAB) is a carcinogen. Handle with extreme care. Avoid breathing vapors. Wear appropriate gloves and safety glasses and use in a chemical fume hood.

1,3-Diaminopropane may be fatal if absorbed through the skin. It is highly toxic and corrosive. The liquid and vapor are both flammable. It is harmful by inhalation, ingestion, or skin absorption. Avoid breathing the dust and vapors. Wear appropriate gloves and safety glasses and use in a chemical fume hood. Keep away from heat, sparks, and open flame.

Dichloromethane (DCM), CH2Cl2 (also known as **Methylene chloride**), is toxic if inhaled, ingested, or absorbed through the skin. It is also an irritant and is suspected to be a carcinogen. Wear appropriate gloves and safety glasses and use in a chemical fume hood. Do not breathe the vapors.

Diethanolamine may be harmful by inhalation, ingestion, or skin absorption. Wear appropriate gloves and safety glasses.

Diethyl pyrocarbonate (DEPC) is a potent protein denaturant and a suspected carcinogen. Aim bottle away from you when opening it; internal pressure can lead to splattering. Wear appropriate gloves, safety gloves, and lab coat and use in a chemical fume hood.

N,N-**Dimethylformamide (DMF), HCON(CH$_3$)$_2$**, is a possible carcinogen and irritating to the eyes, skin, and mucous membranes. It can exert its toxic effects through inhalation, ingestion, or skin absorption. Chronic inhalation can cause liver and kidney damage. Wear appropriate gloves and safety glasses and use in a chemical fume hood.

Dimethyl pimelimidate (DMP) is irritating to the eyes, skin, mucous membranes, and upper respiratory tract. It can exert harmful effects by inhalation, ingestion, or skin absorption. Avoid breathing the vapors. Wear appropriate gloves, face mask, and safety glasses.

Dimethyl sulfoxide (DMSO) may be harmful by inhalation or skin absorption. Wear appropriate gloves and safety glasses and use in a chemical fume hood. DMSO is also combustible. Store in a tightly closed container. Keep away from heat, sparks, and open flame.

Dithiothreitol (DTT) is a strong reducing agent that emits a foul odor. It may be harmful by inhalation, ingestion, or skin absorption. When working with the solid form or highly concentrated stocks, wear appropriate gloves and safety glasses and use in a chemical fume hood.

DMF, *see* **N,N-Dimethylformamide**

DMP, *see* **Dimethyl pimelimidate**

DMSO, *see* **Dimethyl sulfoxide**

DOC, *see* **Deoxycholate**

DTT, *see* **Dithiothreitol**

Dyes, *follow* **manufacturer's safety guidelines**

Epon resin, *see* **Resins**

Ethanol, CH_3CH_2OH, may be harmful by inhalation, ingestion, or skin absorption. Wear appropriate gloves and safety glasses.

Ethanolamine, $HOCH_2CH_2NH_2$, is toxic and harmful by inhalation, ingestion, or skin absorption. Handle with care and avoid any contact with the skin. Wear appropriate gloves and goggles and use in a chemical fume hood. Ethanolamine is highly corrosive and reacts violently with acids.

Ethidium bromide is a powerful mutagen and is toxic. Consult the local institutional safety officer for specific handling and disposal procedures. Avoid breathing the dust. Wear appropriate gloves when working with solutions that contain this dye.

EtOH, *see* **Ethanol**

N-Ethylmorpholine may be harmful by inhalation, ingestion, or skin absorption. Wear appropriate gloves and safety glasses.

Fast Red may cause methemoglobinemia through overexposure. It may be harmful by inhalation, ingestion, or skin absorption. Wear appropriate gloves and safety glasses.

Fixatives, *follow* **manufacturer's safety guidelines**

Formaldehyde, HCHO, is highly toxic and volatile. It is also a carcinogen. It is readily absorbed through the skin and is irritating or destructive to the skin, eyes, mucous membranes, and upper respiratory tract. Avoid breathing the vapors. Wear appropriate gloves and safety glasses and always use in a chemical fume hood. Keep away from heat, sparks, and open flame.

Formamide is teratogenic. The vapor is irritating to the eyes, skin, mucous membranes, and upper respiratory tract. It may be harmful by inhalation, ingestion, or skin absorption. Wear appropriate gloves and safety glasses and always use a chemical fume hood when working with concentrated solutions of formamide. Keep working solutions covered as much as possible.

Formic acid, HCOOH, is highly toxic and extremely destructive to tissue of the mucous membranes, upper respiratory tract, eyes, and skin. It may be harmful by inhalation, ingestion, or skin absorption. Wear appropriate gloves and safety glasses (or face shield) and use in a chemical fume hood.

Glassware, pressurized, must be used with extreme caution. Autoclave and cool sealed bottles in metal containers, pressurize bottles behind Plexiglas shields, and encase 20-liter bottles in wire mesh. Handle glassware under vacuum, such as desiccators, vacuum traps, drying equipment, or a reactor for working under argon atmosphere, with appropriate caution. Always wear safety glasses.

β-Glucuronidase (GUS) may be harmful by inhalation, ingestion, or skin absorption. Wear respirator, appropriate gloves, and safety glasses.

Glutaraldehyde is toxic. It is readily absorbed through the skin and is irritating or destructive to the skin, eyes, mucous membranes, and upper respiratory tract. Wear appropriate gloves and safety glasses and always use in a chemical fume hood.

Glycine may be harmful by inhalation, ingestion, or skin absorption. Wear gloves and safety glasses. Avoid breathing the dust.

Guanidine hydrochloride is irritating to the mucous membranes, upper respiratory tract, skin, and eyes. It may be harmful by inhalation, ingestion, or skin absorption. Wear appropriate gloves and safety glasses. Avoid breathing the dust.

Guanidine thiocyanate may be harmful by inhalation, ingestion, or skin absorption. Wear appropriate gloves and safety glasses.

Guanidinium hydrochloride, *see* **Guanidine hydrochloride**

Guanidinium thiocyanate, *see* Guanidine thiocyanate

HCHO, *see* Formaldehyde

HCl, *see* Hydrochloric acid

$HCON(CH_3)_2$, *see* N,N-Dimethylformamide

$HOCH_2CH_2SH$, *see* β-Mercaptoethanol

H_2SO_4, *see* Sulfuric acid

Hydrochloric acid, HCl, is volatile and may be fatal if inhaled, ingested, or absorbed through the skin. It is extremely destructive to mucous membranes, upper respiratory tract, eyes, and skin. Wear appropriate gloves and safety glasses and use with great care in a chemical fume hood. Wear goggles when handling large quantities.

Hydroxylamine, H_2NOH, is corrosive and toxic. It may be harmful by inhalation, ingestion, or skin absorption. Wear appropriate gloves and safety glasses and use only in a chemical fume hood.

Imidazole is corrosive and may be harmful by inhalation, ingestion, or skin absorption. Wear appropriate gloves and safety glasses and use in a chemical fume hood.

Iminodiacetic acid is an irritant and may be harmful by inhalation, ingestion, or skin absorption. Wear appropriate gloves and safety goggles. Do not breathe the dust.

IPTG, *see* Isopropyl-β-D-thiogalactopyranoside

Isopropanol is flammable and irritating. It may be harmful by inhalation, ingestion, or skin absorption. Wear appropriate gloves and safety glasses. Do not breathe the vapor. Keep away from heat, sparks, and open flame.

Isopropyl-β-D-thiogalactopyranoside (IPTG) may be harmful by inhalation, ingestion, or skin absorption. Wear appropriate gloves and safety glasses.

Kanamycin may be harmful by inhalation, ingestion, or skin absorption. Wear appropriate gloves and safety glasses. Use only in a well-ventilated area.

KCl, *see* Potassium chloride

$K_4Fe(CN)_6$, *see* Potassium ferricyanide

$K_4Fe(CN)_6 \cdot 3H_2O$, *see* Potassium ferrocyanide

$K_2HPO_4/KH_2PO_4/K_3PO_4$, *see* Potassium phosphate

KOH, *see* **Potassium hydroxide**

Lactic acid is corrosive and causes severe irritation and burns to any area of contact. It may be harmful by inhalation, ingestion, or skin absorption. Wear appropriate gloves and safety goggles. Do not breathe vapor or mist.

Leupeptin (or its **hemisulfate**) may be harmful by inhalation, ingestion, or skin absorption. Wear appropriate gloves and safety glasses and use in a chemical fume hood.

LiCl, *see* **Lithium chloride**

Lithium chloride, LiCl, is an irritant to the eyes, skin, mucous membranes, and upper respiratory tract. It may be harmful by inhalation, ingestion, or skin absorption. Wear appropriate gloves, safety goggles, and use in a chemical fume hood. Do not breathe the dust.

Lysozyme is caustic to mucous membranes. Wear appropriate gloves and safety glasses.

Magnesium acetate may be harmful by inhalation, ingestion, or skin absorption. Wear appropriate gloves and safety glasses.

Magnesium chloride, MgCl$_2$, may be harmful by inhalation, ingestion, or skin absorption. Wear appropriate gloves and safety glasses and use in a chemical fume hood.

Magnesium sulfate, MgSO$_4$, may be harmful by inhalation, ingestion, or skin absorption. Wear appropriate gloves and safety glasses, and use in a chemical fume hood.

Maleic acid is toxic and harmful by inhalation, ingestion, or skin absorption. Reaction with water or moist air can release toxic, corrosive, or flammable gases. Do not breathe the vapors or dust. Wear appropriate gloves and safety glasses.

Manganese acetate may be harmful by inhalation, ingestion, or skin absorption. The effects may be delayed. Men exposed to the dusts showed a decrease in fertility. Wear appropriate gloves and safety glasses and use in a chemical fume hood. Do not breathe the dust. Keep away from heat, sparks, and open flame.

Manganese chloride, MnCl$_2$, may be harmful by inhalation, ingestion, or skin absorption. Wear appropriate gloves and safety glasses and use in a chemical fume hood.

MeOH or H$_3$COH, *see* **Methanol**

β-Mercaptoethanol (2-Mercaptoethanol), HOCH$_2$CH$_2$SH, may be fatal if inhaled or absorbed through the skin and is harmful if ingested. High concentrations are extremely destructive to the mucous membranes, upper respiratory tract, skin, and eyes. β-Mercaptoethanol has a very foul odor. Wear appropriate gloves and safety glasses and always use in a chemical fume hood.

MES, *see* 2-(*N*-Morpholino)ethanesulfonic acid

Methanol, MeOH or H$_3$COH, is toxic and can cause blindness. It may be harmful by inhalation, ingestion, or skin absorption. Adequate ventilation is necessary to limit exposure to vapors. Avoid inhaling these vapors. Wear appropriate gloves and goggles and use only in a chemical fume hood.

Methylene blue is irritating to the eyes and skin. It may be harmful by inhalation, ingestion, or skin absorption. Wear appropriate gloves and safety glasses.

Methylene chloride, *see* Dichloromethane

1-Methylpiperazine is highly flammable, combustible, and causes burns. It may be harmful by inhalation, ingestion, or skin absorption. Wear appropriate gloves and safety glasses. Do not breathe the dust. Keep away from heat, sparks, and open flame.

MgCl$_2$, *see* Magnesium chloride

MgSO$_4$, *see* Magnesium sulfate

Mitomycin C is a carcinogen. It may be fatal by inhalation, ingestion, or skin absorption. Do not breathe the dust. Wear appropriate gloves and safety goggles and use only in a chemical fume hood.

MOPS, *see* 3-(*N*-Morpholino)-propanesulfonic acid

2-(*N*-Morpholino)ethanesulfonic acid (MES) may be harmful by inhalation, ingestion, or skin absorption. Wear appropriate gloves and safety glasses.

3-(*N*-Morpholino)-propanesulfonic acid (MOPS) may be harmful by inhalation, ingestion, or skin absorption. It is irritating to mucous membranes and the upper respiratory tract. Wear appropriate gloves and safety glasses and use in a chemical fume hood.

MnCl$_2$, *see* Manganese chloride

NaBH$_4$, *see* Sodium borohydride

Na$_2$CO$_3$, *see* Sodium carbonate

NaF, *see* Sodium fluoride

Na_2HPO_4, *see* **Sodium hydrogen phosphate**

$NaH_2PO_4/Na_2HPO_4/Na_3PO_4$, *see* **Sodium phosphate**

NaN_3, *see* **Sodium azide**

NaOAc, *see* **Sodium acetate** and **Acetic acid**

NaOH, *see* **Sodium hydroxide**

Naphthyl phosphate may be harmful by inhalation, ingestion, or skin absorption. Wear appropriate gloves and safety glasses.

Na_3VO_4, *see* **Sodium orthovanadate**

NH_4HCO_3, *see* **Ammonium bicarbonate**

$(NH_4)_2SO_4$, *see* **Ammonium sulfate**

$(NH_4)_2S_2O_8$, *see* **Ammonium persulfate**

Nickel chloride, $NiCl_2$, is toxic and may be harmful by inhalation, ingestion, or skin absorption. Do not breathe the dust. Wear appropriate gloves and safety glasses.

$NiCl_2$, *see* **Nickel chloride**

Osmium tetroxide (osmic acid), OsO_4, is highly toxic if inhaled, ingested, or absorbed through the skin. Vapors can react with corneal tissues and cause blindness. There is a possible risk of irreversible effects. Wear appropriate gloves and safety goggles and always use in a chemical fume hood. Do not breathe the vapors.

OsO_4, *see* **Osmium tetroxide**

Paraformaldehyde is highly toxic and may be fatal. It may be a carcinogen. It is readily absorbed through the skin and is extremely destructive to the skin, eyes, mucous membranes, and upper respiratory tract. Avoid breathing the dust or vapor. Wear appropriate gloves and safety glasses and use in a chemical fume hood. Keep away from heat, sparks, and open flame.

PEG, *see* **Polyethyleneglycol**

Pentobarbital sodium is toxic and may be harmful by inhalation, ingestion, or skin absorption. It can induce respiratory depression and sedation and presents a risk to the unborn child. Do not breathe the dust. Wear appropriate gloves and safety glasses and use in a chemical fume hood.

Pepsin may be harmful by inhalation, ingestion, or skin absorption. Wear appropriate gloves and safety glasses.

Pepstatin A may be harmful by inhalation, ingestion, or skin absorption. Wear appropriate gloves and safety glasses. Use in a chemical fume hood.

Phenol is extremely toxic, highly corrosive, and can cause severe burns. It may be harmful by inhalation, ingestion, or skin absorption. Wear appropriate gloves, goggles, and protective clothing and always use in a chemical fume hood. Rinse any areas of skin that come in contact with phenol with a large volume of water and wash with soap and water; do not use ethanol!

Phenol red may be harmful by inhalation, ingestion, or skin absorption. Wear appropriate gloves and safety glasses and use in a chemical fume hood.

Phenylenediamine may be harmful by inhalation, ingestion, or skin absorption. Wear appropriate gloves and safety glasses. Use in a chemical fume hood.

Phenylmethylsulfonyl fluoride (PMSF), $C_7H_7FO_2S$, is a highly toxic cholinesterase inhibitor. It is extremely destructive to the mucous membranes of the respiratory tract, eyes, and skin. It may be fatal by inhalation, ingestion, or skin absorption. Wear appropriate gloves and safety glasses and always use in a chemical fume hood. In case of contact, immediately flush eyes or skin with copious amounts of water, and discard contaminated clothing.

Piperidine is highly toxic and is corrosive to the eyes, skin, respiratory tract, and gastrointestinal tract. It reacts violently with acids and oxidizing agents and may be harmful by inhalation, ingestion, or skin absorption. Do not breathe the vapors. Keep away from heat, sparks, and open flame. Wear appropriate gloves and safety glasses and use in a chemical fume hood.

PMSF, *see* **Phenylmethylsulfonyl fluoride**

Polyethyleneglycol (PEG) may be harmful by inhalation, ingestion, or skin absorption. Wear appropriate gloves and safety glasses. Do not breathe the vapor.

Polyvinylpyrrolidone (PVP) may be harmful by inhalation, ingestion, or skin absorption. Wear appropriate gloves and safety glasses and use in a chemical fume hood.

Potassium carbonate may be harmful by inhalation, ingestion, or skin absorption. Wear appropriate gloves and safety glasses and use in a chemical fume hood.

Potassium chloride, KCl, may be harmful by inhalation, ingestion, or skin absorption. Wear appropriate gloves and safety glasses.

Potassium ferricyanide, $K_3Fe(CN)_6$, may be fatal by inhalation, ingestion, or skin absorption. Wear appropriate gloves and safety glasses and always use with extreme care in a chemical fume hood. Keep away from strong acids.

Potassium ferrocyanide, $K_4Fe(CN)_6 \cdot 3H_2O$, may be fatal by inhalation, ingestion, or skin absorption. Wear appropriate gloves and safety glasses and always use with extreme care in a chemical fume hood. Keep away from strong acids.

Potassium hydroxide, KOH and **KOH/methanol,** is highly toxic and may be fatal if swallowed. It may be harmful by inhalation, ingestion, or skin absorption. Solutions are corrosive and can cause severe burns. It should be handled with great care. Wear appropriate gloves and safety goggles.

Potassium phosphate, K_2HPO_4/KH_2PO_4, may be harmful by inhalation, ingestion, or skin absorption. Wear appropriate gloves and safety glasses. Do not breathe the dust. *$K_2HPO_4 \cdot 3H_2O$ is dibasic and KH2PO4 is monobasic.*

PVP, *see* **Polyvinylpyrrolidone**

Pyridine is highly toxic and extremely destructive to the mucous membranes, upper respiratory tract, skin, and eyes. It may be harmful by inhalation, ingestion, or skin absorption. It is a possible mutagen and may cause male infertility. Keep away from heat, sparks, and open flame. Wear appropriate gloves and safety glasses and always use in a chemical fume hood.

Radioactive substances: When planning an experiment that involves the use of radioactivity, include the physicochemical properties of the isotope (half-life, emission type, and energy), the chemical form of the radioactivity, its radioactive concentration (specific activity), total amount, and its chemical concentration. Order and use only as much as is really needed. Always wear appropriate gloves, lab coat, and safety goggles when handling radioactive material. **X rays** and **gamma rays** are electromagnetic waves of very short wavelengths either generated by technical devices or emitted by radioactive materials. They may be emitted isotropically from the source or may be focused into a beam. Their potential dangers depend on the time period of exposure, the intensity experienced, and the wavelengths used. Be aware that appropriate shielding is usually of lead or other similar material. The thickness of the shielding is determined by the energy(s) of the X rays or gamma rays. Consult the local safety office for further guidance in the appropriate use and disposal of radioactive materials. Always monitor thoroughly after using radioisotopes. A convenient calculator to perform routine radioactivity calculations can be found at http://graphpad.com/quickcalcs/index.cfm/.

Resins are suspected carcinogens. The unpolymerized components and dusts may cause toxic reactions including contact allergies with long-term exposure. Avoid breathing the vapors and dusts. Wear appropriate gloves and safety goggles and always use in a chemical fume hood. Sensitivity to these chemicals may develop with repeated contact.

SDS, *see* **Sodium dodecyl sulfate**

Sodium acetate, *see* **Acetic acid**

Sodium azide, NaN₃, is highly poisonous. It blocks the cytochrome electron transport system. Solutions containing sodium azide should be clearly marked. It may be harmful by inhalation, ingestion, or skin absorption. Wear appropriate gloves and safety goggles and handle with great care. Sodium azide is an oxidizing agent and should not be stored near flammable chemicals.

Sodium borohydride, NaBH₄, is corrosive and causes burns. It may be harmful by inhalation, ingestion, or skin absorption. Wear appropriate gloves and safety goggles and use in a chemical fume hood.

Sodium carbonate, Na₂CO₃, may be harmful by inhalation, ingestion, or skin absorption. Wear appropriate gloves and safety glasses and use in a chemical fume hood.

Sodium citrate, *see* **Citric acid**

Sodium deoxycholate is irritating to mucous membranes and the respiratory tract and may be harmful by inhalation, ingestion, or skin absorption. Wear appropriate gloves and safety glasses when handling the powder. Do not breathe the dust.

Sodium dodecyl sulfate (SDS) is toxic, an irritant, and poses a risk of severe damage to the eyes. It may be harmful by inhalation, ingestion, or skin absorption. Wear appropriate gloves and safety goggles. Do not breathe the dust.

Sodium fluoride, NaF, is highly toxic and causes severe irritation. It may be fatal by inhalation, ingestion, or skin absorption. Wear appropriate gloves and safety glasses and use only in a chemical fume hood.

Sodium hydrogen phosphate, Na₂HPO₄ (sodium phosphate, dibasic), may be harmful by inhalation, ingestion, or skin absorption. Wear appropriate gloves and safety glasses and use in a chemical fume hood.

Sodium hydroxide, NaOH, and **solutions containing NaOH** are highly toxic and caustic and should be handled with great care. Wear appropriate gloves and a face mask. All other concentrated bases should be handled in a similar manner.

Sodium lauroylsarcosinate may be harmful by inhalation, ingestion, or skin absorption. Wear appropriate gloves and safety glasses. Do not breathe the dust.

Sodium orthovanadate, Na₃VO₄, may be harmful by inhalation, ingestion, or skin absorption. Wear appropriate gloves and safety glasses and use in a chemical fume hood.

Sodium phosphate, NaH$_2$PO$_4$/Na$_2$HPO$_4$/Na$_3$PO$_4$, is an irritant to the eyes and skin. It may be harmful by inhalation, ingestion, or skin absorption. Wear appropriate gloves and safety goggles. Do not breathe the dust.

Sodium pyrophosphate is an irritant and may be harmful by inhalation, ingestion, or skin absorption. Wear appropriate gloves and safety glasses. Do not breathe the dust.

Sodium vanadate is toxic and is harmful by inhalation, ingestion, or skin absorption. Wear appropriate gloves and safety glasses and use in a chemical fume hood. Do not breathe the dust. Avoid prolonged or repeated exposure.

Streptomycin is toxic and a suspected carcinogen and mutagen. It may cause allergic reactions. It may be harmful by inhalation, ingestion, or skin absorption. Wear appropriate gloves and safety glasses.

Sulfuric acid, H$_2$SO$_4$, is highly toxic and extremely destructive to tissue of the mucous membranes and upper respiratory tract, eyes, and skin. It causes burns, and contact with other materials (e.g., paper) may cause fire. Wear appropriate gloves, safety glasses, and lab coat and use in a chemical fume hood.

TBE, *see* **Tetrabromoethane**

TCEP, *see* **Tris(carboxyethyl) phosphine hydrochloride**

TEMED, *see N,N,N´,N´*-**Tetramethylethylenediamine**

Tetrabromoethane may be fatal by inhalation, ingestion, or skin absorption. It may affect the liver, kidneys, and central nervous system. Wear appropriate gloves and safety goggles and use in a chemical fume hood.

N,N,N´,N´-**Tetramethylethylenediamine** (TEMED) is highly caustic to the eyes and mucous membranes and may be harmful by inhalation, ingestion, or skin absorption. Wear appropriate gloves and tightly sealed safety goggles.

Tetrasodium pyrophosphate, *see* **Sodium pyrophosphate**

Thiourea may be carcinogenic and may be harmful by inhalation, ingestion, or skin absorption. Wear appropriate gloves and safety glasses and use in a chemical fume hood.

Triethanolamine may be harmful by inhalation, ingestion, or skin absorption. Wear appropriate gloves and safety glasses and use only in a chemical fume hood.

Triethylamine is highly toxic and flammable. It is extremely corrosive to the mucous membranes, upper respiratory tract, eyes, and skin. It may be harmful by inhalation, ingestion, or skin absorption. Wear appropriate gloves and safety glasses and use in a chemical fume hood. Keep away from heat, sparks, and open flame.

Trifluoroacetic acid (**TFA**) (**concentrated**) may be harmful by inhalation, ingestion, or skin absorption. Concentrated acids must be handled with great care. Decomposition causes toxic fumes. Wear appropriate gloves and a face mask and use in a chemical fume hood.

Triisobutylsilane may be harmful by inhalation, ingestion, or skin absorption. Wear appropriate gloves and safety glasses. Do not breathe the dust.

Trimethoprim may be harmful by inhalation, ingestion, or skin absorption. Wear appropriate gloves and safety glasses and use in a chemical fume hood.

Trimethylamine *N*-oxide (**TMAO**) causes eye irritation and may be harmful by inhalation, ingestion, or skin absorption. Wear appropriate gloves and safety goggles.

Tris may be harmful by inhalation, ingestion, or skin absorption. Wear appropriate gloves and safety glasses.

Tris (carboxyethyl) phosphine hydrochloride (**TCEP**) is corrosive to the mucous membranes, upper respiratory tract, eyes, and skin and can cause burns. It may be harmful by inhalation, ingestion, or skin absorption. Wear appropriate gloves and safety glasses and use in a chemical fume hood. Do not breathe the vapor or mist.

Triton X-100 causes severe eye irritation and burns. It may be harmful by inhalation, ingestion, or skin absorption. Wear appropriate gloves and safety goggles. Do not breathe the vapor.

Urea may be harmful by inhalation, ingestion, or skin absorption. Wear appropriate gloves and safety glasses.

Urethane is a mutagen and suspected carcinogen. It is also highly toxic and is readily absorbed through the skin. It is be harmful by inhalation, ingestion, or skin absorption. Wear appropriate gloves and safety glasses. Do not breathe the dust and use only in a chemical fume hood.

Xylene is flammable and may be narcotic at high concentrations. It may be harmful by inhalation, ingestion, or skin absorption. Wear appropriate gloves and safety glasses and use only in a chemical fume hood. Keep away from heat, sparks, and open flame.

Xylene cyanol, *see* **Xylene**

Zinc sulfate, ZnSO$_4$, may be harmful by inhalation, ingestion, or skin absorption. Wear appropriate gloves and safety glasses.

ZnSO$_4$, *see* **Zinc sulfate**

Zymolase may be harmful by inhalation, ingestion, or skin absorption. Wear appropriate gloves and safety glasses.

Index

239